GENETIC ENGINEERING OF CROP PLANTS FOR RESISTANCE TO PESTS AND DISEASES

GENETIC ENGINEERING OF CROP PLANTS FOR RESISTANCE TO PESTS AND DISEASES

Edited by W S Pierpoint, *IACR - Rothamsted*
and P R Shewry, *IACR - Long Ashton*

BCPC Registered Office:
49 Downing Street
Farnham
Surrey GU9 7PH, UK.

BRITISH
CROP
PROTECTION
COUNCIL

British Library Cataloguing in Publication Data.
A catalogue record for this book is available from the British Library.

British Crop Protection Council.
Genetic Engineering of Crop Plants for Resistance to Pests and Diseases.

ISBN 0 948404 97 3

Cover design by Major Design & Production Ltd, Nottingham
Printed in Great Britain by Major Print Ltd, Nottingham

CONTRIBUTORS

IACR-Rothamsted, Harpenden, Herts, AL5 2JQ

P.R. Burrows	Entomology and Nematology Department
I.F. Henderson	Entomology and Nematology Department
P.R. Hirsch	Soil Science Department
K.J.D. Hughes	Biochemistry and Physiology Department
P.A. Lazzeri	Biochemistry and Physiology Department
W.S. Pierpoint	Biochemistry and Physiology Department

IACR-Long Ashton Research Station,
Department of Agricultural Sciences, University of Bristol,
Long Ashton, Bristol BS18 9AF

D.M. Glen	Crop and Environmental Sciences Department
J.A. Hargreaves	Cell Biology Department
P.R. Shewry	Director

MAFF Central Science Laboratory,
Hatching Green, Harpenden, Herts, AL5 2BD

J.G. Elphinstone	Bacteriology Department
D.E. Stead	Bacteriology Department

CONTENTS

4. MODIFYING RESISTANCE TO PLANT VIRUSES   **16**

W.S. Pierpoint

5. MODIFYING RESISTANCE TO PLANT-PARASITIC NEMATODES   **38**

P.R. Burrows

1.   INTRODUCTION

Recent advances in the genetic manipulation of plants have greatly increased the ability of crop breeders to produce crop cultivars with desirable characteristics. New techniques offer many new opportunities to develop crops with higher potential yields, crops that better withstand the climatic and chemical stresses of the environment, crops that produce grains or fruit with either better nutritional value or ripening characteristics and, especially, crops with enhanced resistance to pests and pathogens. These are major causes of losses in yield and quality in arable and horticultural crops grown in north-western Europe. In order to combat them a wide range of agrochemicals are used, costing in 1994 in the UK alone, some £200m; the corresponding cost in the European Union is almost eight times greater. The development of natural or engineered mechanisms of crop resistance would clearly have immense economic benefits to the arable and horticultural industries in these countries, while also reducing public concern over pesticide residues in crop products and the environmental consequences of excessive or incorrect pesticide use.

Transformation systems for all the major arable crops and many horticultural crops now allow the development of novel forms of resistances, based either on the manipulation of endogenous plant genes and pathways or the insertion of heterologous genes from other organisms. There has been much conjecture about which harmful agents might be susceptible to control by this approach, so in 1993 the Ministry of Agriculture, Food and Fisheries (MAFF) in the UK, commissioned a report to review these opportunities and to formulate recommendations for future research in this area. The topic was advertised and awarded to IACR (The Institute of Arable Crops Research), in conjunction with MAFF's own Central Science Laboratory. These three research sites, IACR-Rothamsted, IACR-Long Ashton and CSL in Harpenden contain scientists actively engaged in all the relevant disciplines, and who are capable of making objective surveys of specific fields.

The original report was completed in early 1995, using information from three sources. The first was surveys of scientific literature and patents, supplemented by information from talks and poster communications at recent scientific meetings. The second was from informal discussions with active scientists, based in the UK, Western Europe and other parts of the world. The third was questionnaires which were sent out widely to scientists who were actively engaged in relevant research, or who had expressed views on the feasibility and prospects of genetically manipulated crop plants. The questionnaire sought their views on this feasibility, asked their opinion on which pathogens/crops were prime targets for research, and by what means they thought resistance could best be achieved. It also sought their views on the time scale of developments and of the value and acceptability of any produced resistance. Almost 50 detailed replies were received, and the opinions they expressed were used in writing the report and drawing conclusions.

The British Crop Protection Council (BCPC) has long had a strong interest in this important area of agricultural science. Several sessions on aspects of molecular biology have been held at its annual Brighton Conference and its publication list already includes the proceedings of a symposium on *"Opportunities for Molecular Biology in Crop*

Protection", held in 1993. The Council's current policies include giving priority to the integration of biotechnological inputs to crop protection in its conferences, symposia and publications, contributing to debates between researchers and policy makers on biotechnological matters, and collaborating with others to inform the public of the benefits and risks of the application of biotechnology to crop protection. The publication of this monograph, which provides an up-to-date review of the targets for, and approaches to, devising new mechanisms to combat crop pests and diseases, is therefore a timely reinforcement of BCPC's interest.

The scientific section of the original 1994 Report has been revised for publication so that it covers more recent developments. It offers invaluable guidance to researchers, planners, funding agencies and consultants in both the private and public sectors on the perceived prospects for improving crop resistance and environmental protection through genetic engineering, and it indicates where research effort and resources might best be directed.

2.   MAJOR CROP TARGETS

2.1   BACKGROUND

The incidence of different pests and pathogens is influenced by a range of factors, including geographical location, soil type, aspect, year-to-year variation in climate and agronomic practice (e.g. cropping sequence, tillage). Consequently it is impossible to rank the various organisms in terms of relative importance. The major pests and pathogens affecting arable crops in the UK and North-West Europe are therefore listed below with no attempt at ranking.

2.2   MAJOR PESTS AND PATHOGENS THAT AFFECT ARABLE CROPS GROWN IN NORTH-WEST EUROPE

CROP	FUNGI	VIRUSES	BACTERIA	INVERTEBRATES
CEREALS:				
WHEAT Leaves	*Septoria* spp	Barley yellow dwarf virus (BYDV)		Aphids, especially *Sitobion avenae* *Metopolophium* spp *Rhopalosiphum padi* as vectors of BYDV and *as pests
	Puccinia spp: rusts			
	Erysiphe graminis f.sp. *tritici*: mildew			
	Cephalosporium gramineum wheat leaf stripe			
	Drechslera tritici repentis tan spot			
Roots & Stem bases	*Gaeumannomyces graminis*: take-all			*Delia coarctata* Wheat bulb fly
	Pseudocercosporella herpotrichoides: eyespot			Leatherjackets *Tipula* spp
	Fusarium spp (also on ears)			
	Rhizoctonia cerealis sharp eyespot			Slugs *Deroceras reticulatum* *Arion* spp

3

Ears	*Tilletia caries* bunt		Wheat blossom midges *Sitodiplosismosellana* *Contarinia tritici*
	Ustilago nuda loose smut		
	Claviceps purpurea ergot		

BARLEY

Leaves	*Erysiphe graminis* f.sp. *hordei*: mildew	Barley yellow dwarf virus (BYDV)	Aphids as vectors of BYDV
	Rhynchosporium secalis: leaf blotch	Barley yellow mosaic virus (BaYMV)	
	Pyrenophora teres: net blotch	Barley mild mosaic virus (BaMMV)	
	Puccinia spp: rusts		
	Selenophoma		
Roots & Stem Bases	*Pseudocercosporella herpotrichoides*: eyespot		Slugs esp *Deroceras reticulatum* *Arion* spp
	Fusarium spp		
	Gaeumannomyces graminis: take all		Nematodes esp *Heterodera avenae*
	Polymyxa spp as vectors of BaYMV and BaMMV		
	Rhizoctonia cerealis: sharp eyespot		
Ears	*Ustilago nuda* loose smut		
	Claviceps purpurea ergot		

OATS

Leaves	*Puccinia coronata* crown rust	Barley yellow dwarf virus (BYDV)	*Oscinella frit*: Frit fly

	Erysiphe graminis mildew		Aphids as vectors by BYDV

BRASSICA SPP.

OILSEED RAPE and other Brassicas	*Leptosphaeria maculans* canker	Beet western yellows virus (BWYV)	Insects *Ceutorhynchus assimilis* cabbage seed weevil
	Pyrenopeziza brassicae: light leaf spot	Cauliflower mosaic virus (CaMV)	*Dasineura brassicae* Brassica pod midge
			Psylliodes chrysocephala Cabbage stem flea beetle
	Alternaria brassicae: dark leaf and pod spot	Turnip mosaic virus (TuMV)	*Meligethes aeneus* Bronze blossom beetle
	Sclerotinia sclerotiorum: stem rot		*Myzus persicae* peach-potato aphid as vector of BWYV and TuMV
	Peronospora parasitica: downy mildew		*Brevicoryne brassicae* Cabbage aphid
	Mycosphaerella capsellae: white leaf spot		*Pieris* spp Cabbage white catterpillar
	Plasmodiophora brassicae: club root		Slugs esp *Deroceras reticulatum* *Arion* spp

SUGAR BEET

			Aphids esp *Myzus persicae* as vector of BYV, BMYV and BMV
	Erysiphe betae: powdery mildew	Beet mild yellowing virus (BMYV)	
	Polymyxa betae: vector for BNYVV rhizomania	Beet yellows virus (BYV)	*Pegomya hyoscyami* Mangold fly
	Uromyces betae: leaf rust	Beet necrotic yellow vein virus (BNYVV)	*Tipula paludosa* Leatherjackets
	Helicobasidium brebissonia violet root rot	Beet mosaic virus (BtMV)	Cyst nematode *Heterodera schachtii*
			Slugs *Deroceras reticulatum* *Arion* spp

POTATOES

Phytophthora infestans: late blight	Potato leaf roll virus (PLRV)	*Erwinia carotovora* sub sp *atroseptica*: black leg and soft spot	Aphids esp *Myzus persicae* as vectors of PVLR and PVY
Spongospora subterranea: powdery scab also as vector of PMTV	Potato virus Y (PVY) and related Potyviruses	*Erwinia carotovora* sub sp *carotovora*: soft rot in store	Nematodes *Globodera rostochiensis* *G. pallida*
Rhizoctonia solani: stem canker and black scurf	Potato mop-top virus (PMTV) Tobacco rattle virus (TRV)		Slugs *Deroceras reticulatum* *Arion* spp *Tandonia* spp
Helminthosporium solani: silver scurf			
Polyscytalum pustulans: skin spot			

LINSEED

Botrytis cinerea: grey mould		Flea beetles esp *Aphthona euphorbiae* *Longitarsus parvulus*
Alternaria linicola		*Thrips angusticeps*
Oidium lini: powdery mildew		
Septoria linicola: pasmo		
Verticillium dahliae		
Sclerotinia sclerotiorum		

LEGUMES
Peas

Peronospora viciae: downy mildew	Pea enation mosaic virus (PEMV)	*Psendomonas syringae*: bacterial blight	*Sitona lineatus*: pea weevil
Mycosphaerella pinodes	Pea seed-borne mosaic virus (PSMV)		Thrips *Thrips angusticeps* *Kakothrips robustus*
Botrytis spp			*Acyrthosiphon pisum* pea aphid vector of PEMV
Fusarium spp foot rot			
			Cydia nigricana:

	Sclerotinia sclerotiorum		pea moth
Beans	*Ascochyta fabae*	Bean leaf roll virus (BLRV)	*Aphis fabae*: black bean aphid also vector of BLRV and PEMV
	Uromyces fabae: rust		
	Peronospora viciae: downy mildew	Pea enation mosaic virus (PEMV)	*Bruchus rufimanus*: bean seed beetle
	Botrytis spp		Nematode *Ditylenchus dipsaci*

2.3   CONCLUSIONS

Three groups of organisms include important pests of all the major arable crops, namely fungi, insects either as pests and/or viral vectors, and viruses. Slugs and nematodes are important pests of specific crops, notably cereals, oilseed rape, sugar beet and potatoes. Bacterial pathogens are only important pests of potatoes, where they cause rots of field and stored crops.

7

3.   ENABLING TECHNOLOGY FOR NORTH-WEST EUROPEAN CROPS

P.A. Lazzeri
(IACR-Rothamsted)

3.1   BACKGROUND

Two established technologies for the transformation of plants by foreign genes exist, the use of the Ti plasmid of the soil bacterium *Agrobacterium tumefaciens* as a biological vector for foreign gene transfer (Ooms, 1992), and the use of "direct gene transfer" methodology, in which physical methods are used to transfer foreign DNA into plant cells (Potrykus, 1990). Several different methods for direct gene transfer have been developed, including protoplast transformation, cell or tissue electroporation, vortexing tissues with silicon carbide fibres and the bombardment of tissues with DNA-coated tungsten or gold particles. The latter technique, "particle bombardment", is the most successful and widely applied (Klein *et al.*, 1992).

At present, transformation methods rely on the capacity to regenerate plants from tissue cultures of the target plant, and the lack of robust *in vitro* regeneration systems is, in a number of species, a limiting factor in the application of genetic engineering techniques. Although plant regeneration has been achieved from *in vitro* cultures of all of the major temperate crops, in a number of important plants the efficiency and reproducibility of the regeneration process is genotype-dependent and a limited number of cultivars can be handled routinely.

Of the two transformation technologies, the mechanism of *Agrobacterium*-mediated transformation is better understood, and more is known about the characteristics of the transgenic plants recovered (Hooykaas and Schilperoort, 1992). In most cases, a single or low number (1 - 4) copies of a transgene will be integrated at a single genetic locus and this gene(s) will be meiotically stable and transmitted to progeny as a single dominant locus (there are exceptions to this rule; the implications for crop plant genetic engineering are discussed in **3.4** below). Direct gene transfer is a less controlled process and may result in more complicated patterns of transgene integration, with the insertion of multiple and/or rearranged copies of the transgene. This can lead to a higher frequency of transgene instability and/or deviations from normal Mendelian segregation ratios (see **3.4** below).

For these reasons, *Agrobacterium* - mediated transformation is generally the preferred method for crop plant genetic engineering, with particle bombardment being employed for species intractable to *Agrobacterium*. The applicability of *Agrobacterium* - mediated transformation is largely decided by the natural host-range of the bacterium, but the mode of plant regeneration is a second critical factor, as *Agrobacterium* only interacts efficiently with wounded cells. This means that crop plants which typically regenerate in culture via organogenesis (shoot formation) (e.g. *Solanaceous* species) are generally more amenable to *Agrobacterium* transformation than species which typically regenerate via somatic embryogenesis (e.g. most legumes).

3.2   TRANSFORMATION STATUS OF MAJOR CROPS; LIMITATIONS AND PROSPECTS

3.2.1   Cereals

Cereals are recalcitrant to *Agrobacterium* infection and their transformation only became really feasible with the advent of the particle gun. Worldwide, fertile transgenic plants of wheat, barley, oats and triticale have all been produced in the last four years, but the transformation of rye has not yet been published (Lazzeri and Shewry, 1993). Within Europe, a number of academic and industrial laboratories work on the transformation of wheat and barley. Several laboratories have recovered transgenic wheat lines (e.g. at IACR-Rothamsted and Nickersons/Biocem in the UK and at the University of Hamburg in Germany) but there has been slower progress in barley transformation. Some groups have produced barley transformants (e.g. the University of Hamburg, Carlsberg Laboratories in Denmark) but transformation is still not routine. There is comparatively less work on oat transformation; one centre is IGER-Aberystwyth in the UK, but to date there is no published report of transgenic oat plants in Europe. There is relatively little work on rye or triticale, although both of these species have been transformed. Most current cereal work is based on genotypes selected for good response *in vitro*, but adequate regeneration can now be obtained from elite cultivars and the plants recovered from the culture systems used show little variation and good fertility. With present levels of effort on wheat and barley transformation, reliable procedures applicable to elite cultivars can be expected within 3-4 years.

3.2.2   Forage Grasses

Among temperate fodder grasses, fescue species have been transformed, but there has been slower progress with the economically important ryegrass species (*Lolium perenne* and *L. multiflorum*). The centre of research in the UK is IGER-Aberystwyth where transgenic tall fescue (*Festuca arundinacea*) plants have been recovered (Dalton and Bettany, 1994), and similar work is in progress by groups in the Netherlands and Switzerland who have also recovered transformants. Good culture systems exist for the ryegrass species, so that the prospects for transformation by the bombardment of embryogenic tissues in the next 1-3 years are good.

3.2.3   Potato

Potatoes are amenable to *Agrobacterium*-mediated transformation, so that transgenic plants are produced routinely by many groups worldwide, including a number of academic and industrial laboratories throughout Europe. There is still significant cultivar-dependent variation in transformation efficiency, but this is not a major limitation to the engineering of pest or pathogen resistance traits as the majority of cultivars respond at adequate levels. Field trials of transgenic potatoes have already been performed in a number of European countries.

3.2.4   Sugar Beet

This plant is difficult to manipulate *in vitro*, and although a number of laboratories

9

worldwide have produced transgenic plants (via *Agrobacterium* transformation), procedures are slow and inefficient. Most expertise resides in industrial groups associated with seed or agrochemical companies. In sugar beet, the lack of a facile transformation procedure is at present a limitation to applied genetic manipulation, particularly within the public sector. However, a novel transformation procedure based on targeting specific leaf cells has recently been developed from joint Dutch/British/ Belgian research and this may lead to significant increases in efficiency.

3.2.5   Oilseed Rape, *Brassica* spp.

Among *Brassica* species, OSR transformation is most developed and is routine in a number of European laboratories. There are still deficiencies in terms of reproducibility and genotypic variation, but the majority of cultivars can be handled. Due to the importance of OSR as a European crop and the relatively clear opportunities for the application of genetic engineering to alter lipid qualities, most major agrochemical and seed companies are active in OSR transformation and a significant number of field trials have been performed. Other economically important *Brassicas* (*B.oleracea*, *B.campestris*, *B.rapa*, *B.juncea* etc) have been transformed, but systems are generally less well developed than for OSR.

3.2.6   Grain Legumes

Worldwide, this major economic group of plants has proved difficult to manipulate. Culture systems exist, but these are not readily-amenable either to *Agrobacterium* infection or bombardment. Nevertheless, with the exception of *Vicia faba* and *Lupinus alba* the major grain legumes have now been transformed, albeit at low efficiency and in a few genotypes only (Christou, 1994). Of the grain legumes important throughout Northern Europe, pea is the only crop which has been transformed, but the technology is at present inefficient and reproducible in only a few laboratories. At present transformation capabilities certainly limit the feasibility of engineering modified grain legume crops.

3.2.7   Forage Legumes

Small-seed legumes are generally easier to manipulate *in vitro* than grain legumes, but robust transformation techniques have still proved difficult to develop (Christou, 1994). *Trifolium repens*, *Lotus corniculatus* and *Medicago sativa* have all been transformed, but there are still limitations in terms of reproducibility and genotype-dependence, particularly for *T.repens*. There is active research on forage legume manipulation in the UK, France, Netherlands and Denmark, both in academic and industrial laboratories. Current transformation capabilities for *L.corniculatus* and *M.sativa* are probably adequate, but clover transformation is in need of development to allow crop improvement via genetic engineering.

3.2.8   Tomato, Salad Crops, (Cucurbits, lettuce, celery, chicory etc.), Carrot

As other *solanaceous* crops, tomato is amenable to transformation by *Agrobacterium*, although genotypic variation limits efficiency in some cultivars. Many academic laboratories use tomato as a model for genetic manipulation work and the crop is handled widely in industrial laboratories. The first food product from engineered tomatoes (tomato paste from plants with modified ripening due to altered ethylene synthesis, from Zeneca Plc) will appear in European markets in 1996. Most other salad crops and carrot have been transformed (*Agrobacterium*), and there is some activity in industrial laboratories (particularly in the Netherlands) and Universities. In general, transformation techniques are available for these species and worldwide there is considerable experience, but it is also the case that if specific cultivars are to be targeted for genetic manipulation then improvement/adaptation of procedures may be required.

3.2.9   Fibre Crops (Flax, hemp)

Worldwide, a number of laboratories are active in flax transformation (*Agrobacterium* - mediated), but there is limited activity in Europe, although interest will increase if the culture of the crop increases and targets for manipulation are identified. Little published information is available on hemp transformation.

3.2.10   Soft Fruit Crops

In vitro regeneration technology has been available for some time for most of these crops, and major species (e.g. apple, pear, plum, cherry, strawberry) have been transformed by *Agrobacterium* procedures. Most species are, however, complicated to handle because of factors such as juvenility/long fruiting cycles and slow growth rates. In addition, efficient techniques exist for only few elite clones. Centres of activity throughout Europe include Britain, France and Belgium. For most soft fruit crops the current status is that disease resistance-engineering is feasible in only a few elite cultivars and further development of transformation technology will be needed if wider application is envisaged.

3.2.11   Trees

Some species, including poplars and willows, are relatively amenable to manipulation *in vitro* and to *Agrobacterium* transformation, but the majority of woody species are difficult to culture and to transform. Progress has been slower than in non-woody species because relatively few research groups are active worldwide. Gymnosperms are typically not readily susceptible to *Agrobacterium* transformation but efficient somatic embryogenesis systems exist for several coniferous species, allowing transformation by particle bombardment (Van Doorsselaere *et al.*, 1993). There are laboratories active in woody plant manipulation throughout Europe, but as these species are generally less important economically than field crops, the total effort is still relatively modest.

3.2.12   Ornamental Species

Worldwide, there is considerable interest in genetic engineering of ornamentals and in Europe, Dutch breeding/biotechnology companies are particularly active. Some ornamentals (e.g. *Petunia*) are simple to manipulate *in vitro* and are used as model species, while others such as rose have until recently proved recalcitrant, despite major efforts. In general, most progress has been made with dicotyledonous species. Monocots, such as many of the common bulbs, show low susceptibility to *Agrobacterium* infection, and bombardment-mediated transformation procedures are only recently being developed.

3.3   AVAILABILITY OF MARKER GENES AND REGULATORY SEQUENCES

3.3.1   Marker Genes

Over the past few years a range of different selectable and/or scorable marker genes and their respective selection agents/substrates have been assessed in many different plant transformation systems. From this testing a general consensus has emerged in that most workers use one of two selectable marker systems, either aminoglycoside antibiotic selection, using the neomycinphosphotransferase (NPT II/*neo*) resistance gene, or the herbicide resistance gene phosphinothricin acetyltransferase (PAT/*bar*) which allows selection with the herbicide Basta (or the closely-related chemicals Bialophos or Challenge). As a scorable marker, the β-glucuronidase (GUS) gene is used almost universally for the histological assessment of transgene expression. The only important deficiency in terms of marker genes for plant transformation is the lack of a good vital marker. The firefly luciferase gene has been available for some years but it requires sophisticated detection equipment for its use. An alternative system which may have potential for use in plants is jellyfish "green fluorescent protein" (GFP) (Chalfie *et al.*, 1994); this would be a technological advance of major significance.

3.3.2   Regulatory Sequences

While adequate selectable marker systems for crop transformation are available the choice of regulatory sequences (promoters) for the control of transgene expression is still somewhat limited and there is as yet little good comparative data on the performance of those promoters which are available. For the engineering of agronomic traits such as pest or disease resistance in crop plants, it is clear that there is the need for spatial and temporal regulation of transgene expression to ensure that resistance genes are expressed at the right time and in the right parts of the plant to effect protection. In many cases it will be desirable to have inducible defence gene expression in response to pest or pathogen attack. Constitutive transgene expression has the disadvantages that it may aid the development of resistance among pathogens and high levels of constitutive expression may have negative consequences for the plant and be undesirable if the protein is present in harvested plant parts. Although there is currently considerable effort towards the identification, characterisation and isolation of regulatory sequences for defence gene expression, at present most of the model experiments (and nearly all field studies) on engineered resistance have used constitutive promoters, most often the cauliflower mosaic CaMV35S sequence. A primary aim should therefore be to identify and test regulatory sequences for defence gene expression. In some cases it may be acceptable to use heterologous promoters from another plant (or alternative source) to direct expression, but in many cases more precise control of expression will be required and here homologous sequences may need to be isolated from the plant to be protected, or from a close relative. For example, a nematode-induced, root-specific

12

promoter isolated from *Arabidopsis* may well not show the same function in wheat, necessitating the isolation of a homologous gene from wheat (if this exists) or the isolation of an appropriate cereal-specific sequence. A further consideration is that plant-derived promoters are likely to be more acceptable to regulatory authorities than promoters derived from pathogens such as viruses.

3.4 STABILITY AND HERITABILITY OF TRANSGENE EXPRESSION, AND FIELD TRIALS OF TRANSGENIC PLANTS

At the end of 1992 there had already been some 770 field releases of transgenic crops world wide, 227 of which were in Europe (Beck and Ulrich, 1993) and the 1993 and 1994 seasons will probably have added some 50 further trials to the European total. This implies that a number of transgenic crops have undergone considerable assessment and selection at the hands of breeders to produce suitable material for field trials. Unfortunately, most of the data on transgene heritability, segregation patterns and the stability of expression over generations which must have been amassed, is not in the public domain. Information on the stability of expression and transmission of transgenes and on potential environmental influences on transgene expression is clearly of central importance to the practical application of genetic engineering but this is an area which is only starting to be studied in depth. It is becoming clear that there may be interactions both between transgenes and between transgenes and endogenous plant genes which may result in gene inactivation and may also lead to heritable changes in gene expression (Finnegan and McElroy, 1994). The empirical approach to such phenomena, and the one applied to date, is to produce populations of transformants and, over time, to select those individuals having the required patterns and stability of expression. However, it appears that transgene instability is associated with particular patterns of integration, both in *Agrobacterium-* and direct gene transfer transformants (Flavell, 1994), and with a better understanding of the mechanisms involved we can expect to develop means of stabilizing transgene expression.

A further aspect of transgene stability and one which comes to the fore in consideration of field experiments, is the potential for the transfer of transgenes to weedy relatives of crop species or for "horizontal" transfer to other organisms. While this topic is not within the scope of the present survey, this is an area of some public concern and potential hazards associated with the release of transgenic plants must clearly be assessed rigorously. There is, however, no *a priori* reason why transgenes should be transferred at any higher frequency than genes introduced into crops by conventional genetic means, so much "risk assessment" can be done using conventional genetic markers rather than transgenes themselves. One class of transgenic plants do raise particular cause for concern; these are plants in which genes from pathogens, particularly viruses, are expressed systemically. In such cases there might be the potential for a second infecting virus to recombine with the engineered molecule to produce a novel pathogenic genome with new characteristics (see Section 4.5). Such manipulations must obviously be scrutinised in great detail.

3.5 CONCLUSIONS

Transformation methods exist for all of the major Northern European crops which are likely to be targets for the genetic engineering of resistance, but in a number of crops significant improvements in efficiency and in applicability to a range of cultivars are required.

Adequate selectable and scorable marker systems are available, although a better vital marker for gene expression would speed the development of more efficient transformation procedures. There is need also for the identification and testing of more regulatory sequences allowing inducible, tissue-specific and developmentally-regulated transgene expression. Different sets of promoters may be needed for different crop groups.

There is the need for more field trials with transgenic plants, to amass more knowledge on the performance of regulatory sequences in the field environment and on the stability and heritability of transgene expression.

3.6   REFERENCES

Beck, C.I. & Ulrich, T.H. (1993) Environmental release permits. *Bio/Technology*, **11**, 1524-1528.

Chalfie, M., Tu, Y., Euskirchen, G., Ward, W.W. & Prasher, D.C. (1994) Green fluorescent protein as a marker for gene expression. *Science*, **263**, 802-805.

Christou, P. (1994) Genetic engineering of crop legumes and cereals: current status and recent advances. *Agro-Food-Industry Hi-Tech*, **5**, 17-27.

Dalton, S. & Bettany, A. (1994) AFRC News, January 1994, p5.

Finnegan, J. & McElroy, D. (1994) Transgene inactivation: plants fight back! *Bio/Technology*, **12**, 883-888.

Flavell, R.B. (1994) Inactivation of gene expression in plants as a consequence of specific sequence duplication. *Proceedings of the National Academy of Sciences of the USA*, **91**, 3490-3496.

Hooykaas, P.J.J. & Schilperoort, R.A. (1992) *Agrobacterium* and plant genetic engineering. *Plant Molecular Biology*, **19**, 15-38.

Klein, T.M., Arentzen, R., Lewis, P.A. & Fitzpatrick-McElligott S. (1992) Transformation of microbes, plants and animals by particle bombardment. *Plant Physiology*, **91**, 440-444.

Lazzeri, P.A. & Shewry, P.R. (1993) Biotechnology of cereals. *Biotechnology and Genetic Engineering Reviews*, **11**, 79-146.

Ooms, G. (1992) Genetic engineering of plants and cultures. In *Plant Biotechnology* (M.W. Fowler and G.S. Warren, Eds.) pp. 223-257. Pergamon Press, Oxford.

Potrykus, I. (1990) Gene transfer to cereals: an assessment. *Bio/Technology*, **8**, 535-542.

Van Doorsselaere, J., Baucher, M., Van der Mijnsbrugge, K., Leple, J.C., Rohde, A., Van Montagu, M. & Inze, D. (1993) Genetic engineering in forest trees. *Agro-Food-Industry Hi-Tech*, **6**, 15-19.

4.   MODIFYING RESISTANCE TO PLANT VIRUSES

W.S. Pierpoint
(IACR-Rothamsted)

4.1   BACKGROUND

The advent of genetic manipulation of plants and of viruses has transformed the pace and vitality of research into virus-host relationships. Not only has it produced some crop plants with enhanced virus resistance, but it has revealed new and unexpected mechanisms of resistance that are as yet not understood, and which almost certainly have very far reaching implications. It is currently unclear whether these mechanisms are an expression of "natural" resistance mechanisms or are in some way specifically induced by introduced genes or gene fragments. Opportunities for creative research in this area are currently very high, and have recently been generally reviewed by Harrison (1992), Wilson (1993) and Baulcombe (1994b) among others. The aspects considered here are divided into natural resistance mechanisms, resistance based on genes derived from viral genomes,and resistance derived from genes for toxins and non-plant proteins. As so little is known of the mechanisms by which resistance is achieved, these divisions, and especially their subdivisions, may prove to be quite arbitrary.

4.2   INVESTIGATION AND EXPLOITATION OF NATURAL RESISTANCE MECHANISMS

4.2.1   Major Resistance Genes

A major resistance gene, specifically the N-gene of tobacco that confers a "hypersensitive" resistance to tobacco mosaic virus (TMV), has been cloned and sequenced for the first time, and its structure briefly reported (Baker, see Moffat, 1994). Such genes, usually specific to a particular virus or virus strain, have been introduced into many crop species by conventional breeding techniques and in many cases have provided important, durable resistances. The prospect of being able to transfer these genes at will into the genomes of established but unrelated crop species is extremely attractive, and there are many research projects aimed at isolating them. Thus, the Tm-2 gene that confers resistance to tomato mosaic virus on tomato plants, and the potato Rx gene which confers resistance against potato virus X (PVX) are being located and cloned by research groups in Cornell University (USA) and the John Innes Centre: it seems likely that the structures of these and other similar genes will be available in 2 or 3 years time. The tobacco N gene was located and isolated with the use of the AC transposon from maize; its structure has only recently been published in detail, and it appears to contain some features which are common to isolated antifungal R-genes (see section 8.2). These features include a region containing repeats of a leucine-rich motif that may be involved in protein - protein interactions, a P-loop that suggests a nucleotide - binding area and a region with some similarities to a protein kinase. There is no obvious transmembrane region, and first impressions suggest a large, cytoplasmic, membrane - anchored protein, suited to recognise intracellular challenges and to respond to these through recognisable cell-signal transducing systems. The successful transformation of other plant species with a gene of this size and type lies in the future; but its functional integration into the metabolic machinery of a recipient plant cell clearly depends upon the presence of the complex systems with which it interacts.

The structural features of invading viruses that are recognised by such resistance genes, and so trigger the resistance response, are poorly understood. With structurally simple plant viruses it is often assumed that these are located on the coat protein, and there is evidence that it is, for example, the coat protein of TMV that interacts with the hypersensitive N' gene of *Nicotiana sylvestris* (Culver *et al.*, 1994). But recognition may be different and more complex with other virus-gene combinations. Thus the coat protein of PVX is probably important as an initial recognition signal that interacts with a potato resistance-gene, Rx, but other features of the virus are involved in subsequent reactions that restrict virus accumulation (Baulcombe *et al.*, 1994). Moreover, the feature that involves resistance that is conferred by the Nb gene, another important potato gene, probably resides not in a protein but in the RNA sequence of PVX that codes for the viral RNA polymerase: changes in this sequence that do not affect its coding instructions nor its translatability, nevertheless prevent it acting as an avirulence determinant (Baulcombe *et al.*, 1994).

4.2.2   Pathogenesis - Related and other Resistance Induced Proteins

A common expression of the resistance induced by an anti-viral R or N gene is a hypersensitive response (HR) in which virus spread and multiplication are restricted to a small, often necrotic area centred around the infection site. This response is often associated with a heightened resistance to a second infection with the virus or indeed to many other pathogens which evoke an HR: moreover this acquired resistance is manifest not only in infected leaves, but in other unaffected leaves of the plant, the so-called systemic acquired resistance (SAR). Many studies have revealed the biochemical complexity of this HR response and identified many novel substances that are produced during its course (see Fritig *et al.*, 1987), but they have not identified those that are directly responsible for restricting virus multiplication or spread. Indeed, it is often difficult to devise test systems that could suitably detect and assess such critical activities. This has encouraged the production of transgenic plants, usually tobacco, which express some of these novel components, and which can be tested directly for increased resistance to pathogens. Attention has been mainly focused on the Pathogenesis - Related proteins (PR's) of tobacco (see van Loon 1985), a class of proteins that are an obvious and major component of the HR in this and other species and whose genes can now be readily identified and cloned (see Linthorst 1991). Moreover there were good reasons for believing the proteins to be involved with resistance (e.g. Bol *et al.*, 1990): some of them belong to a group of 9 families of proteins whose genes have been specifically associated with systemically acquired resistance (Ward *et al.*, 1991).

So far however, judging from published reports, the genes for PR-proteins that have been introduced into tobacco plants, have not conferred any obvious virus resistance. These genes include those for some of the PR-1 class of proteins, PR-1a and 1b, that for a PR-5, thaumatin-like protein , and that for a glycine-rich protein, GRP, that has not been identified but which is closely associated with the HR (see Bol *et al.,* 1990; Alexander *et al.,* 1993). They have been constitutively expressed both in tobacco cultivars containing the N-gene and those without it, but had no effect either on systemically spreading infections nor on localised, avirulent ones. It can be argued that these proteins are only antiviral when produced in specific combinations. Testing this possibility for all the five established families of PR-proteins, each of which contains both intercellular acidic as well as intracellular basic isoforms which in turn are coded for by distinct families of genes, is a laborious and expensive operation. It has, however, been undertaken in a thorough and systematic way by a research group of the Ciba-Geigy Corporation (mainly centred in N Carolina, USA), and all the transformed plants are screened for resistance to fungal and bacterial pathogens, as well as to viruses (see Alexander *et al.,* 1993). The emphasis of the screening is now especially orientated towards fungal pathogens (see section 8.2), since three classes of PR proteins, the PR-2s (glucanases), PR-3's (chitinases), and PR-5's (thaumatin-like proteins), are known to have antifungal activities, which often act synergistically, in *in vitro* tests. An unexpected result of this research programme, is that tobacco plants expressing a PR-1 are resistant to two fungal (oomycete) pathogens which cause the black shank and blue mold diseases. No function, antifungal or otherwise, had previously been ascribed to PR-1 proteins, which are made in large amounts during the HR of tobacco, although they have been the object of much discussion.

4.2.3   Resistance Mediated by Satellite Nucleic Acids

The severity of disease symptoms produced by a small number of plant viruses can be markedly affected, either increased or decreased, by the presence of species of small "satellite" nucleic acids. These are typically small pieces of RNA which are apparently unrelated to the virus genome: they do not invariably accompany their 'helper virus', but can only replicate in its presence when they may become encapsidated in its coat protein (CP) (see Harrison, 1992 for refs). The best known example is the satellite RNAs which accompany cucumber mosaic virus (CMV). Their ability to affect the host plants tolerance to CMV has been known for a long time, and for over 10 years selected benign strains of the satellite have been used commercially in China to protect pepper and other crops from the severe effects of CMV (Tien *et al.,* 1987; Harrison, 1992). This protective effect of benign satellites to a specific virus is, happily, retained when plants are transformed so as to produce the satellite RNA constitutively (i.e. without infection).

In the first test of this method of producing virus tolerance, tobacco plants were transformed with the DNA equivalent of 1.3-2.3 copies of the RNA satellite of CMV (Harrison, 1992). They produced little transcript RNA until challenged with CMV, when large amounts of RNA were produced and packaged in CMV-like particles. The plants were as susceptible to infection as were untransformed controls, and virus replicated freely in the inoculated leaves. However, replication and the severity of symptoms were largely suppressed in systemically infected and newly developing leaves. Also, as Harrison stresses, these recovering plants are very poor sources of virus for vector aphids, so that virus transmission from them is poor and those plants which do become infected are virtually symptomless; clearly the attenuating satellite is aphid transmitted along with the virus. Tobacco plants can also be transformed to produce a similar resistance to the unrelated tobacco ringspot nepovirus (TRSV: see Harrison, 1992), Harrison (1992) summarizes the properties of this satellite-induced tolerance and contrasts it with that

18

conferred by transgenically produced viral coat protein (CP). He concludes that it is likely to be durable in field conditions, and although it is probably not effective against even viruses closely related to the helper virus, it can be combined in the same plant with the broader resistance mediated by CP against the same virus (Harrison & Murant, 1989). Wilson (1993) takes a less sanguine view of its long term prospects, mainly because of the risk of a single mutation altering the beneficial characteristics of a specific transduced satellite RNA.

The mechanism by which benign satellite RNAs protect a plant is unknown. They may inhibit the replication of helper viruses, and in some cases ameliorate symptoms by preventing virus CP from entering and disrupting chloroplasts (see Wilson, 1993). Their effect on virus replication has been compared by Baulcombe (1994b) to "decoy molecules, distracting proteins away from the inoculated virus genome and therefore away from involvement in virus replication and spread". In this and in other respects also, they have been compared with the defective-interfering (DI) nucleic acids that occur naturally with two groups of plant viruses (Tombusviruses and Carmoviruses) and many animal viruses. These molecules are rearranged fragments of the helper virus genome, but can, like satellite RNAs, be packaged in virus CP and can affect, for better or worse, the severity of infection.

Transformed plants which produce DI nucleic acids, like those transformed to produce satellite RNAs, are less affected by the appropriate helper virus. *Nicotiana benthamiana* for instance, can be made less sensitive to the DNA-containing, geminivirus, African cassava mosaic virus (ACMV), so that upon inoculation it shows less symptoms and supports less virus replication (Stanley *et al.*, 1990): moreover, both protective effects are enhanced following serial infection between transformed plants. More recently, the same plant species was made markedly resistant to the lethal apical necrosis caused by the RNA virus *Cymbidium* ringspot virus (CyRSV) by a similar tranformation (Kollar *et al.*, 1993). This resistance was not overcome by high concentrations of inoculum nor by infectious viral RNA, and it was not correlated with the degree of transgene transcription in the transformed plants. These examples have encouraged attempts to extend the range of viruses to which protection of this type might be genetically engineered, by creating artificial DI nucleic acids based upon specific viral genomes (see Baulcombe, 1994b). Where these are effective however, it is not clear if the DI-nucleic acids are acting as 'decoy molecules' or if they are examples of the RNA-mediated resistance mechanisms reviewed by Baulcombe (1994a) and mentioned below.

4.3   PATHOGEN GENOME - DERIVED RESISTANCE (PGDR)

4.3.1   Coat Protein - Mediated Resistance

Coat protein - mediated resistance (CPMR) can arguably be regarded as an extension of a well studied form of natural resistance, that of cross-protection. In this phenomenon, infection of a plant with a 'mild' or attenuated strain of a virus confers upon it a resistance to subsequent infection by more virulent strains of the virus and even of other, closely related viruses. Cross-protection has been known for many years and used not only as a means of assessing the relationship of virus isolates, but as a practical way of protecting field crops of tomatoes, apples and citrus fruits. Among the many speculations as to the mechanism by which this protection works, was the suggestion that the coat protein (CP) of the protecting virus was somehow involved. This suggestion received substantial support in 1986, when Powell-Abel *et al.*, showed that transgenic tobacco plants, expressing the coat protein gene from TMV, were resistant to subsequent infection by this virus. This phenomenon has been demonstrated with other plant-virus combinations, and in 1992 Harrison could list examples involving viruses from 10 taxonomic groups. The phenotypic characteristics of this type of resistance are that infection occurs at fewer sites, that there is a reduced rate of systemic disease development through the plant and usually, a decreased accumulation of virus. Other general characteristics (Beachy *et al.*, 1990: Harrison *et al.*, 1992: Wilson 1993) are that resistance can be overcome by high concentrations of inoculum, and that it does not protect against infection by viral RNA although it may extend to intact virions of closely related strains or viruses. An important characteristic of the early examples of CPMR is that it depends upon, and may be proportional to, the degree of expression of the introduced CP-gene. It should be stressed that this summary contains a degree of generalisation and not all these characteristics apply to all the CPMR cases listed by Harrison (1992) and Beachy *et al.*, (1990). Moreover the phenomenon merges confusingly into types of resistance that can be conferred by transformation with pieces of virus genome other than CP, and as one reviewer (Wilson, 1993) comments, "The vast literature (on CPMR) reveals many details unique to each virus-plant-CP system and even some patterns common to several viruses, but recent cases add more exceptions than rules...". Nevertheless the phenomenon seems currently to have enough individual characteristics to consider it separately from other examples of PGDR.

Both Harrison (1992) and Wilson (1993) commented that lists of the first, now 'classic' cases of CPMR concerned only viruses with simple, positive-sense ssRNA genomes, and that all the transformed plants were dicots. The expanding list of new cases removes both these limitations. Thus the monocot rice has been transformed with the CP-gene of rice stripe virus (RSV) to be resistant to this planthopper-borne virus. RSV is a pathogen of economic importance and whose genome contains both ds- and ss- RNA. The transformed plants were genetically stable, produced CP in amounts up to 0.5% of their total soluble protein and had a virus resistance which appeared to depend on CP production (Hayakawa *et al.*, 1992). The range of viruses susceptible to CPMR was extended when the transferred CP-gene of a DNA geminivirus was shown to confer resistance to tomato plants (Kunik *et al.*, 1994). The virus, tomato yellow leaf curl virus (TYLCV), has a genome consisting of a single circular strand of DNA, and is whitefly transmitted. Resistance is manifest as a delay in the development of the disease followed by a systemic recovery that produces new leaves that are resistant to subsequent infection: again resistance appeared restricted to plants producing CP.

The mechanism of CPMR is not clear. It seems certain that in many cases it involves the inhibition of a very early stage in virus disassembly and translation (see Harrison, 1992: Wilson, 1993), but it almost certainly affects other stages of replication and transport. But one mechanism may not serve to explain all examples of what may legitimately be described as CPMR. Thus tobacco plants can be transformed with parts of, or with the whole of the nucleocapsid protein gene from a strain of tomato spotted wilt virus (TSWV) to make them resistant to this virus and to closely related strains. In many transformed lines (e.g. Haan *et al.*, 1992; Kim *et al.*, 1994), this specific resistance is not mediated by expressed CP : but Pang *et al.*,(1993) observed that some transformed plants expressing high levels of this protein were resistant to virus strains that have a low homology with the gene-donor strain and even to other, distantly related tospoviruses. It can be argued with some conviction that this resistance is due to the "heteroencapsidation" of the RNA of the challenging viruses by constitutively produced nucleocapsid protein so as to produce a dysfunctional particle that somehow interferes with viral replication.

The first field trials of CP-modified tomatoes and potatoes involved deliberate virus - inoculation and gave encouraging results (Harrison, 1992). They have been continued, and it is indeed likely that the bulk of the 50 or so field trials of transgenic plants with resistant traits that took place world wide up to 1991 (see Wilson, 1993), involved CPMR. Van den Elzen *et al.*, (1993) concluded from 4 years of field trials that "the engineering of commercial crops [of potato] against PVX or Fusarium has demonstrated unequivocally the potential of genetic engineering technology". A level of PVX field resistance was obtained that was equivalent to resistance conferred by a classical vertical resistance gene. Moreover, after suitable selection, lines were obtained that preserved the intrinsic properties of the plant cultivar. This is of course, especially important for heterozygous polyploid crops like potato whose quality traits are critical in industrial food processing. A careful selection was also necessary in other field trials of transformed potatoes to obtain greatest resistance to PVX and potato virus (PVY). Performance of lines from individual transformations could not be predicted from growth chamber tests nor from the degree of expression of the CP gene. It is possible that the highly resistant lines finally selected in such a programme, do not always owe their resistance to the presence of CP itself.

It is widely believed that there are now many CP-transformed plants that are near to marketing; examples include the potatoes resistant to both PVY and PVX developed by the Monsanto Company (USA), and yellow squash resistant to zuccini yellow mosaic virus (Shah *et al.*,1995). It may be that such transgenic plants are already used practically in some parts of the world (Harrison, 1992); and indeed, it is estimated that almost 30% of the tobacco currently produced in China, the product of some 6 million hectares of land, is produced from transgenic plants (Dale, 1995; Plafker, 1994). Recent reference to field trials and to CPMR research are provided in a review by Baulcombe (1994b). Baulcombe also discusses situations where CPMR has proved not to be reliable or where it may present a potential hazard. Thus, the resistance of potatoes to tobacco rattle virus does not extend to infection *via* viruliferous nematodes, and that of other plants to cucumber mosaic virus breaks down at high temperatures. Of more concern are situations where transgenically-produced CP can encapsidate the RNA of another virus. It has already been mentioned that such encapsidation may result in resistance to distantly related viruses; but it has been claimed, although not yet documented, that the CP of potato leafroll virus can encapsidate the RNA of viruses or viroids to make them insect-transmissible (see Baulcombe, 1994b). These risks are common to all resistance strategies that induce plants to produce a functional component of a virus. They have to be assessed individually with consideration of the chemistry and biology of the specific virus protein, of the crop plant, and also of the environment in which the transformed plant will be grown.

4.3.2　Resistance Mediated by Replicases and some other Viral Non-structural Proteins

Transgenic tobacco plants containing part of the replicase gene (the RNA-dependent RNA polymerase) of TMV are highly resistant to TMV inoculation (Golemboski et al., 1990). Although it proved impossible to detect the expected protein product of the transgene, mutagenesis experiments led to the belief that this protein was indeed necessary for the resistance. This interesting effect has a ready explanation, similar to the simplest mechanism proposed for CPMR; the replicase protein, whether it is fully functional or not, would be expected to act as a decoy molecule, and to upset the delicate equilibrium among the components of the plants transcription complex, an equilibrium that is necessary for its activity. Since this first example, there have been over half a dozen descriptions of replicase-mediated resistance involving other virus-plant combinations. These are listed and discussed by Wilson (1993) and Baulcombe (1994a) among others. The viruses involved are RNA viruses such as CMV, PVX and PVY and pea early browning virus (PEBV), but Wilson (1993), citing work in progress, finds no reason to believe that the effect will not extend to DNA viruses also. The examples include cases where the transgene extends from a component of the replicase (e.g. PEBV) or to a dysfunctional mutant of it (PVX), to the expression of the intact replicase gene (CyRSV).

Replicase - mediated resistance involves a strong inhibition of virus replication, and can be extremely effective against high levels of inoculum. In tests involving PVX and transgenic tobacco plants, it was judged to be more effective than CPMR (Braun & Hemenway, 1992). It extends to isolated protoplasts and to infective viral RNA. It is, however, also highly strain-specific, and only effective against those virus strains that are closely related to the source of the transgene. Its mechanism is certainly more complex than the simple explanation given above, and may be multiple: it is a matter of active investigation and debate. It is still possible to argue (Baulcombe, 1994a) that in some instances, the replicase or a subunit of it are involved and necessary even if they are difficult to detect. However, in many examples, as is the situation with CPMR, there is no clear relationship between the extent to which the introduced protein is formed and the degree of resistance: in a number of instances, including the resistance of transgenic plants to PVX (see Baulcombe, 1993; Longstaff et al., 1993) the most resistant plants are those showing very low levels of replicase protein synthesis. Baulcombe (1994a) summarises these cases and argues that they are mediated by introducing mRNA rather than by protein, and so they involve a mechanism similar to that discussed in the next section.

Whatever the mechanism(s) of replicase - mediated resistance, its properties suggest that it could form the basis of useful field resistance for crop plants. Recently workers at the Monsanto Company (USA) reported that a resistance against potato leaf roll virus (PLRV) in Russet Burbank potatoes held up very well in two years of field trials (Shah et al., 1994:1995). Its most obvious limitation in this respect is its restricted specificity. Its spectrum of effectiveness could obviously be extended by introducing a number of transgenes to deal with the different virus isolates that can occur in field populations. This rather crude approach will almost certainly be superseded by more sophisticated methods. These however will only come from a deeper knowledge of the resistance mechanism and this is a relatively long term expectation.

Viral genes which code for non-structural proteins other than replicases may also confer virus resistance in transgenic plants. Thus many potyviruses contain a protein which is covalently linked to their RNA, and which acts as a protease during the processing of viral RNA. When tobacco is transformed with the protease gene from tobacco vein-mottling virus (TVMV) it shows a strong, highly specific resistance to TVMV infection (Maiti et al., 1993). This resistance has some resemblance to those mediated by RNA: and although the transformed plants are thought to produce detectable amounts of protease, Baulcombe (1994b) cautions that until constructs are made that produce non-translatable protease transcripts, and these are shown not to produce resistant plants, the role of the protease in resistance will be questionable. This resistance contrasts with the 'broad spectrum' resistance conferred by the CP gene of PVY. But, significantly, the TVMV gene that codes for a cylindrical inclusion protein, appears not to confer resistance to transformed plants (Maiti et al., 1993). Thus there appear to be some limits to pathogen genome-derived resistance and not every piece of virus genome is a potential resistance gene.

4.3.3   Resistance Associated with Movement- and Helper-proteins

Many, and possibly all, plant viruses contain genes whose products are essential for the movement of virus progeny from an infected cell into a neighbouring, uninfected, cell. This process is essential for the systemic invasion of the host, and helps, of course, to determine the virulence of the infection and possibly the host range of the virus. Movement is thought to be via the plasmodesmata, the narrow cytoplasmic threads that pass through cell walls and link adjacent cells. The structure of these threads is such as to restrict the movement between cells of cytoplasmic particles to molecules smaller than 3 nm with molecular weights below 0.7 kD (Citovsky et al., 1992: Deom et al., 1990). These sizes are smaller than those of intact viruses and of their infective components. The viruses are believed to produce specific proteins, the movement proteins (MP) or transport proteins, which can affect the structure and porosity of the plasmodesmata and so facilitate virus spread. Interference with the functioning of these proteins by a plant cell component would be expected to restrict or prevent virus spread and so confer a resistance to systemic infection on the plant (for reviews see Hull, 1989; Deom et al., 1992; Maule, 1991; Lucas & Gilbertson, 1994).

The most convincing evidence for the function of these MPs, and of the possibility of disrupting their function, comes from studies on TMV. This virus codes for a 32 kD protein, which, from mutational studies has long been thought to be involved in cell-to-cell movement. This was confirmed by Deom et al., (1987) who introduced the protein into transgenic Xanthi tobacco, where it facilitated ("complemented") infection by a TMV mutant (Ls1) that is deficient in cell-to-cell movement in a temperature sensitive (ts) manner. At temperatures that would normally be non-permissive to Ls1, chlorotic lesions spread on inoculated leaves and virus appeared systemically in younger, non-inoculated leaves. Moreover the speed of systemic spread of Ls1 that normally takes place at permissive temperatures, was increased in the transformed plants. These effects were only observed in plants actively expressing MP; in such plants MP has been shown to accumulate in cell walls, and the molecular exclusion limits of these plants

23

plasmodesmata, as judged by dye-diffusion techniques, are increased 3-4 fold over those of untransformed plants (Deom et al., 1990; Deom et al., 1992).

In complementary experiments, tobacco plants were transformed with the gene for the MP of another ts TMV mutant, which is also deficient in cell-to-cell movement (Malyshenko et al., 1993). They had, as expected, a ts response to wild type TMV. Although susceptible to TMV at 24°C, when maintained at 33°C, inoculated leaves accumulated less than a tenth as much TMV as did plants not expressing MP. The obvious interpretation of this effect is that,at the higher temperature, the ts MP, while not being completely functional, can still compete in some function with the MP produced by the wild-type TMV, and so interfere with its normal functioning. But a more striking and encouraging demonstration of such an interference in transformed tobacco plants occurred when they expressed TMV-MP which had been deliberately rendered dysfunctional by deleting 3 amino acids from near its N-terminus (Lipidot et al., 1993). This protein is not able to increase the permeability of tobacco plasmodesmata to large molecules, but it apparently interferes with the ability of TMV-MP to do so, so that following inoculation with TMV, virus spread, replication and lesion development are severely delayed. This holds whether the dysfunctional MP is expressed in a susceptible host (Xanthi nn) or one that responds hypersensitively to TMV (Xanthi NN). Appropriate control experiments make it very likely that the basis for these effects of dysfunctional MP are on virus spread and not on ease of infection or on virus replication per se.

It is premature to judge the potential of this form of virus resistance. It is expected to be applicable to many viruses, and also to have a broad specificity. Thus, evidence of different types suggests the presence of genes for MPs in many viruses (e.g. Hull, 1989; Koonin et al., 1990), and at least one other MP, that from alfalfa mosaic virus (AlMV), increases the permeability of the plasmodesmata of transgenic plants (Poirson et al., 1993). Moreover tobacco plants expressing the MP of brome mosaic virus (BMV) have a degree of resistance to the unrelated TMV, as if this MP were acting as a dysfunctional MP for this virus also (Malyshenko et al., 1993). The degree of resistance to infection that a dysfunctional MP will confer, will doubtless be increased when more is known of the cellular interactions of MPs and of their functional domains.

The necessary background on the properties and functioning of MPs, is being sought internationally. It is becoming clear that the MPs of viruses including TMV and cauliflower mosaic virus (CaMV) not only affect plasmodesmata, but also bind to single stranded nucleic acids so as to keep them in long, unfolded configurations (Citovsky et al., 1992; Deom et al., 1992): this is obviously relevant to the ease with which infective particles may pass through plasmodesmata. Just as fundamental, it is now clear that some viruses modify plasmodesmata in a different and characteristic way by extending a tubular cytoplasmic structure through them: the width of these tubes being sufficient to allow movement of intact viral particles (Deom et al., 1992). This modification occurs in infections caused by members of at least six groups of viruses with spherical particles, including cowpea mosaic (CPMV:comovirus), cauliflower mosaic virus (CaMV:caulimovirus) and nepoviruses. The MPs of the viruses are usually larger (~45-50 kDa) than that of TMV. The functional analysis of the MP of CaMV is the subject of active study in the John Innes Institute (A.J.Maule; personal communication).

Another class of virus-coded proteins which mediate an aspect of virus movement are the "helper component" (HC) proteins or "aphid transmission factors" that are essential for virus transmission by homopterous insects, especially aphids. They are presumed to function by interacting with both virus coat protein and some part of the insect vector. Disrupting this two headed attachment would be expected to prevent the insect transmission of a wide range of important pathogens.

Most research on these proteins concerns the HCs of potyviruses. Thus tobacco plants expressing the 50 kD HC of tobacco vein mottling virus (TVMV) have been produced, and the protein shown to be effective in facilitating the transmission of, for example, purified tobacco etch virus (TEV; Berger *et al.*, 1989). The transformation system is less straightforward than that involving MPs, however, mainly because of the mechanism of replication of potyviruses. Their viral genomes are usually translated as a large polyprotein that requires as many as seven proteolytic cleavages to produce the individual gene products. To obtain plants expressing detectable amounts of HC, it was necessary to transform the plants not only with the fragment of genome coding for HC, but with substantial pieces of adjacent genome also. This larger region, the HC-protease region, is known from mutation experiments (Atreya & Pirone, 1993) to affect aspects of virus replication and symptom development as well as aphid transmission. While these complications may make it difficult to interpret the results of resistance studies on plants expressing either HC or dysfunctional HC, they may also prove beneficial as transgenic fragments coding for dysfunctional HC-protease, may interfere with other aspects of virus infection as well as aphid transmissibility.

Unpublished experiments (Hunt & Pirone; personal communication) have detected virus resistance in some plants transformed directly to produce HC. Upon infection the plants show initial symptoms, but later recover. This recovery may resemble that described by Dougherty and colleagues for plants transformed with other potyvirus genes (see below), and be due to a similar mechanism.

Other viruses which are insect transmitted and may depend on HC-type proteins are listed by Hull (1994). They include caulimoviruses, carlaviruses, closteroviruses and a rice virus (rice tungro spherical virus, RTSV) which is transmitted by a leafhopper. The best characterised of their HCs is that of CaMV. It is comparatively small (18 kD) and has recently been expressed in a baculovirus-insect cell system where it accumulates in a paracrystalline form (Blanc *et al.*, 1993). It is the subject of current research at a number of European laboratories.

4.3.4   RNA-Mediated Resistance

An increasingly large number of examples are known where resistance has been conferred on a transgenic plant by the introduction of a piece of viral genome which does not code for a protein. Sometimes protein expression is expected but can not be detected in spite of determined attempts. One of many examples is the resistance of potato plants transformed with the coat protein of PLRV (Barker *et al.*, 1993). Sometimes the introduced gene is incapable of being translated or has deliberately been rendered so before introduction. An interesting example of this type of transformation involving tobacco plants and the potyviruses, tobacco etch virus (TEV) and PVY, has been described by Dougherty and his colleagues (Lindbo & Dougherty, 1992a, 1992b; Lindbo et al., 1993; Silva-Rosales et al., 1994). Plants containing genes for full length TEV-CP, or for the CP truncated at its N-terminus, produced the expected proteins which accumulated in amounts up to 0.01% of soluble leaf protein. They showed a resistance to the virus in that symptoms were attenuated and eventually younger, "recovered" leaves emerged, devoid of both symptoms and virus. However, plants containing a gene for a C-terminal truncated CP, or plants containing genes for CP-antisense or CP genes rendered untranslatable by the inclusion of stop codons, produced, as expected, no new proteins. Nevertheless they were highly resistant to TEV infection as were protoplasts derived from them. This resistance was highly specific for the gene-donor strain of TEV, was not overcome by high levels of inoculum, and was active against aphid-borne virus. Thus two different types of resistance appeared to have been induced by different gene constructs, although surprisingly, plants containing full length transcripts of CP, and which have "recovered" from TEV-infection, show a highly specific resistance to further infection that may also be mediated by RNA. Initially Lindbo & Dougherty (1992b) interpreted some of their results in terms of the hybridisation of RNA transcripts to RNA replicative intermediates. However, observations on the steady state levels of transgenic RNA in "recovered" tissue led them to propose (Lindbo et al., 1993) that resistance is due to a specific, cytoplasmic RNA - degrading mechanism that is primed by, and whose specificity is determined by, the RNA sequences of the transgene. Such a process, it was pointed out, would resemble in many ways the "co-suppression" phenomenon well known in plants transformed with genes other than virus genes (Jorgensen, 1990). Here, attempts to over-express a particular gene product, by the introduction of an additional gene, often result in the suppression of transcription of both endogenous and introduced gene. Recent experiments with tobacco plants transformed with modified forms of the RNA-polymerase gene from PVX, confirm a relationship between conferred virus resistance, the accumulation of only very low levels of transgenic RNA, and the suppression or "silencing" of homologous, endogenous genes (Mueller et al., 1995); they strongly support the idea of the induction of a specific RNA-degrading mechanism.

The earliest attempts to use antisense genes deliberately targeted at virus components so as to produce resistant plants, had only limited success. It was anticipated that the gene transcripts would hybridize with viral RNA coding for the CP's of CMV, PVX and TMV, and so prevent viral replication. These results have been briefly reviewed by Wilson (1993), who argued that this approach might be expected to be less effective against high copy number RNA viruses which have a cytoplasmic replicative cycle, and perhaps more successful with viruses of the gemini- and caulimovirus groups whose replication involves a nuclear phase. However this approach has been re-orientated and revitalized by the reports of virus-resistance engineered by antisense constructs (as well as sense-RNA constructs) derived from the PVY genome (Lindbo et al., 1992b), as well as similar results involving PLRV (Barker et al., 1994; Kawchuck et al., 1991) and tomato spotted wilt virus (Pang et al., 1993). It becomes an obvious and important research priority to unravel the mechanism(s) that underlie this type of resistance, so that it may be manipulated and exploited in the most advantageous manner.

26

Comparatively little has been reported on the field performance of these newly described, RNA-mediated resistances. That induced against PLRV (assuming that it is RNA-mediated) has been introduced at the Scottish Crop Research Institute into a breeding clone of potatoes that had some degree of host-mediated resistance to PLRV. The resulting high level of resistance to virus-replication, is transmitted satisfactorily to plants developed from infected tubers (Barker *et al.*, 1994). In discussing a similar PLRV resistance in Russet Burbank, the most popular North American potato cultivar, Kawchuk *et al.*, (1991) comment that transgenically produced antisence RNA is less likely to create a possible hazard than is sense RNA, as it is unlikely to be encapsidated by the CP of infecting viruses: 'heteroencapsidation' of sense-RNA is thought to be possible with these insect-transmitted luteoviruses.

A strategy for producing resistance to viruses that is theoretically related to the use of antisense genes, is that of introducing genes for ribozymes. Ribozymes are comparatively small RNA molecules with the potential to cleave catalytically at specific sites in RNA molecules to which they can hybridize. They are natural components of the replication cycle of some viroids and satellite RNA's which involve the auto-processing of large RNA transcripts. Synthetic ribozymes can now be designed to target other pieces of RNA, including viral genes which are essential in virus replication (see Harrison, 1992; Wilson, 1993). Ribozymes have been synthesised which cleave TMV-RNA (Edington *et al.*, 1992; Gerlach *et al.*, 1990) and PLRV-RNA *in vitro* (see Wilson, 1993). The anti-TMV constructs have also been introduced genetically into tobacco plants and tobacco protoplasts where they are reported to inhibit virus replication to a limited degree. A problem with their functioning *in vivo* that has been revealed in other, non-viral studies, is to obtain expression in quantities sufficient for them to "swamp" the target RNA. This has led to the suggestion (see Wilson, 1993) that a better use of ribozymes is to modify mild, attenuated strains of virus so that they produce, as subgenomic RNA's, ribozymes specific for other more damaging viruses. This could be regarded as an effective enhancement of the classical cross-protection phenomenon.

The practical usefulness of antiviral ribozymes has yet to be demonstrated and the comparatively little research that they attract is mainly in the USA and Australia. If substantial protection against viruses is observed in ribozyme transformed plants, it will have to be demonstrated that introduced RNA is acting in a ribozyme-like manner, and not, for example, acting as antisense or by the sort of RNA-scavenging mechanism described above.

4.4 RESISTANCE BASED ON OTHER PROTEINS INCLUDING TOXINS AND NON-PLANT PROTEINS

4.4.1 Plant Produced Antibodies

Transformed plants are capable of synthesizing the protein chains of mammalian antibodies (Abs) and assembling they into fully functional complexes. This has raised the anticipation among plant pathologists that plants may be directed to produce antibodies against specific essential components of pathogens, especially viruses, and by complexing these components, restrict pathogen replication. There is much interest in this possibility, both here and abroad, and in both industrial and academic organizations, but few results have been reported publicly.

Wilson (1993) described briefly some unpublished work, both of his own and others, which used antibodies raised against the CP of TMV and the nucleoprotein of tomato spotted wilt virus (TSWV), and which was aimed at inhibiting the cotranslational disassembly of these viruses that is thought necessary to initiate infection. He comments on some difficulties encountered by himself and others, such as getting Abs expressed at the levels which *in vitro* experiments suggest are necessary, and of getting Abs assembled and accumulated in the appropriate cellular compartment; there is some evidence that they are more effective when expressed extracellularly. He suggests that it may be better to use Abs directed against viral replicative enzymes rather than against structural CPs, as these will be produced in comparatively small amounts. Encouraging results however were reported by Tavladoraki and his colleagues (1993) who produced transgenic *N. benthamiana* plants which produce a single chain (scFv) monoclonal antibody active against the CP of artichoke mottled crinkle tombusvirus (AMCV), and which are resistant to infection with this virus. Following mechanical inoculation of these plants, fewer (~50%)became infected, and of those that did, infection developed slowly and virus accumulation, especially in systematically infected leaves, was very much decreased. It is not yet know at what stage virus replication is inhibited. The authors attribute the success of their experiments to the use of single chain antibodies which are clearly active in the cytoplasm, and which, unlike whole antibody molecules, need no special targeting to the endoplasmic reticulum for the correct folding and assembly processes. This work is likely to provide an encouraging role model for many future experiments. Thus collaborative work between the Scottish Crop Research Institute and the University of Leicester, is exploring the use of scFvs against the potato viruses PVX and PVY.

4.4.2   Toxins and other "Suicide" Genes

A possible and much-discussed way of producing virus-resistant plants, is to introduce a gene for a toxic product which, when activated by virus infection, kills the infected cell so limiting virus replication and spread. The toxin could either be produced constitutively and sequestered relatively harmlessly in the cell until infection, or alternatively be induced specifically following infection. The idea behind this approach can be traced to an old interpretation of the hypersensitive response to infection. The toxins most often considered in this context are the ribosome-inactivating proteins (RIPs), which were recently reviewed by Stirpe *et al.*, (1992).

Plant RIPs are a family of comparatively small (25-32 kD) proteins that occur in leaves, seeds and roots, and which are very effective inhibitors of ribosomal translation. Some 40 specific forms and isoforms have been purified and studied, but it is thought that they may be quite widespread in plants where they play a defensive role against pathogens 39 (Stirpe *et al.*, 1992). Most of them consist of a single peptide chain, and, like the antiviral protein of pokeweed (*Phytolacca americana*) have a limited toxicity to mammals. Others, the type-2 RIPs, contain two peptide chains linked by covalent as well as non-covalent bonds, one of which has a lectin-like bonding domain. These are among the most potent of natural toxins against mammals, the most notorious being ricin from the seeds of castor beans (*Ricinus communis*). In spite of this toxicity, the type-2 RIPs are thought

28

to have an important potential, when targeted towards specific cells, as chemotherapeutic agents. But it is the less toxic, type-1 RIPs which occur in foodstuffs such as cereal grains (Stirpe et al., 1992), that attract attention as plant protectants. Both types of compounds are believed to be effective by cleaving a specific N-glycosidic bond in the 28S ribosomal RNA (Hartley *et al.*, 1991) so preventing the binding of elongation factor 2 and inhibiting protein synthesis. Ribosomes from different organisms, mammals, bacteria and plants, and even from different genera of plants, have however, different sensitivities to different RIPs.

Lodge et al., (1993) transformed tobacco and potato plants with the gene for an RIP from pokeweed which, because it is an inhibitor of the mechanical inoculation of viruses (see Chen et al., 1991), is also known as the pokeweed antiviral protein (PAP). The plants were found to have acquired a resistance to viruses. This work illustrates the difficulties as well as the potential for dealing with this type of toxin. Transformation events, using *Agrobacterium tumefaciens* as transformation vector, were infrequent, and their recovery low. Plants expressing the highest levels of PAP were sterile and had growth abnormalities. Both of these effects were attributed to the intracellular presence of toxin, even though most of it was sequestered near to the cell walls and in intercellular fluids. However, satisfactory plants were produced which contained about 1-5 ng of toxin per mg protein, and they were shown to have resistance against the unrelated viruses potato virus X (PVX), potato virus Y (PVY) and cucumber mosaic virus (CMV). This expected wide range of resistance, which extends to aphid transmission, is, of course, one of the advantages of this type of approach. The adverse effects of the intercellular toxin may well be overcome if the introduced gene were expressed as antisence RNA, controlled by a suitable RNA promoter, so that it were only translated into toxic protein following virus infection (see Wilson, 1993).

Research on the RIPs of pokeweed (Chen *et al.*, 1993) and of *Dianthus* as well as other plants (Taylor *et al.*, 1994) is being done in the UK at IACR (Rothamsted), at Warwick University and at the John Innes Institute. Wilson (1993) mentions other toxins, including diptheria toxin, which are being used in related research, but which are unlikely to be useful in an agricultural context.

4.4.3   Other Proteins

There are several examples where plants transformed with genes which are not obviously related to plant virus infection have been shown to have virus resistance. This occurs, for instance, in tobacco plants expressing a gene for a variant of the protein ubiquitin and which have, in consequence, a low level of ubiquitin itself (Becker *et al.*, 1993). If the plants belong to varieties that respond hypersensitively to TMV, they are less susceptible to inoculation than are controls, and produce fewer lesions. If they are susceptible varieties, TMV replication is inhibited but not abolished. Ubiquitin is an essential component of a cellular process that removes damaged proteins, and it is known to be involved in plant responses to such stresses as heat shock. It could conceivably be involved in the normal responses to virus infection, and indeed there is a little evidence for this belief (see Becker *et al.*, 1993). An important factor in motivating this research was that plants expressing moderate amounts of the variant gene, tend to produce lesions which are superficially similar to virus-induced lesions, when they are subject to mild, abiotic stresses. It is, perhaps, a little less surprising that plants transformed with the gene for a mammalian 2-5 oligoadenylate synthase, a component of the mammalian interferon system, should be tested for, and found to have, resistance to potato virus X (PVX; Truve *et al.*, 1993). However, the relevance of the interferon system to plants has not received unanimous recognition in spite of emphatic claims that exogenously applied interferons and 2-5 oligoadenylates inhibit the replication of plant viruses (see Kulaeva *et al.*, 1992). Truve *et al.*, (1993) believe that virus resistance in the transformed plants involves the activation of a rapid RNA-degrading mechanism. Evidence is needed as to whether this resembles that involved in mammalian cell responses to interferon, or resembles the RNA-mediated systems referred to above.

4.5   RISKS AND CONJECTURAL HAZARDS

Some of the problems presented by the use of crop plants with transgenically-derived resistance to viruses, such as the transference of undesirable genes to wild species, are common to all transgenic plants. However, there are conjectural hazards that are specifically presented by plants containing genes derived from viral genomes. These are considered by a number of reviewers (e.g. Harrison, 1992: Wilson, 1993). They include firstly, the possibility that transgenically introduced CP will encapsidate heterologous viral RNA, and make it, for example, aphid-transmissible: secondly, endogenous, viral-derived RNA may recombine with the RNA of infecting viruses to give new variants: thirdly, tranduced satellite or DI nucleic acids may mutate to give less benign and more damaging forms. All these processes, of course, may occur naturally and some can be demonstrated in the laboratory. Thus a non-aphid transmissible strain of the zucchini yellow mosaic virus (ZYMV) was made aphid-transmissible by passage through a transgenic plant expressing the CP of another, aphid-transmissible potyvirus (Lecoq *et al.*, 1993). While recognizing that these processes may occur in transgenic plants, Harrison (1992) concluded that "there is no evidence that the risks are greater in practice than those posed by conventionally bred cultivars or (those) that occur naturally in other ways". Wilson (1993) emphasized that the use of dysfunctional fragments of viral genome may minimize some of the risks, and that risk assessment is now open to direct experimentation.

Public debate on this issue followed the recent report that a movement-defective strain of cowpea chlorotic mottle virus, defective because of the lack of a third of its capsid gene, was rendered functional on being inoculated into transformed *N. benthamiana* expressing two thirds of this gene including the missing fragment. Pressure groups subsequently called for a moratorium on the commercialization of such transformed plants, pending the introduction of a stronger (US) risk assessment programme (Wuethrich, 1994). However, this type of recombination depends on similarities between the expressed gene and a gene of the invading virus, and also on the size of the regions with suitable similarities as well as on the selection pressure applied. On this basis, Falk & Bruening (1994) have made an assessment of the likelihood of these recombinations occurring in transgenic crops. They conclude that it is unlikely that they "will occur at frequencies greater than they are already occurring by combination between genomic RNAs in natural conventional and subliminal infections". Moreover, they concluded, it is unlikely that any new viruses that may be formed, will be more viable than competing viruses throughout the full infection cycle. Most of the virologists responding to a questionnaire devised in connection with this report, expressed a similar general opinion. Risks connected with the use of plants transformed with fragments of virus genome, were small and manageable. Nevertheless these risks need to be addressed experimentally, and field releases monitored, in order to allay official and public apprehension.

4.6   CONCLUSIONS

The techniques of genetic manipulation have produced transformed crop plants with resistance against specific viruses. Some of these plants have been developed, following field trials and selections, to the point of market release. Many others are being developed or could be so. Especially important are those resistances for which there is no known source of natural resistance which could be used in conventional breeding programmes.

The most developed forms of resistance are those based on fragments of the virus genome, especially those concerned with virus coat proteins and replicases. These two types of resistance differ in a number of respects including their specificity against particular strains of viruses. Resistance based on movement-proteins may, when developed, be of very broad specificity. The mechanisms by which the different forms of resistance are produced are essentially unknown, and in many cases are not what was initially expected. However, genetic manipulation has created exciting research possibilities which will certainly help in elucidating these resistance mechanisms, including those that appear to be mediated by RNA. This information will almost certainly lead to the development of new, more satisfactory forms of resistance, possibly for instance, of very broad specificity.

The risks associated with the use of genes derived from the virus genome are considered to be more perceived than real: nevertheless, they need examining and assessing for the sake of official and public reassurance.

4.7   REFERENCES

Alexander, D., Goodman, R.M., Gutt-Rella, M. et al., (1993) Increased tolerance to two oomycete pathogens in transgenic tobacco expressing pathogenesis-related protein 1a *Proceedings of the National Academy of Sciences of the USA*, **90**, 7327-7331.

Atreya, C.D. & Pirone, T.P. (1993) Mutational analysis of the helper-component-proteinase gene of a potyvirus : effects of amino acid substitutions, deletions , and gene replacement on virulence and aphid transmissibility. *Proceedings of the National Academy of Sciences of the USA*, **90**, 11919-11923

Barker, H., Reavy, B., Webster, K.D. et al., (1993). Relationship between transcript production and virus resistance in transgenic tobacco expressing the PLRV coat protein gene. *Plant Cell Reports*, **13**, 54-58

Barker, H., Reavy, B., Arif, M. *et al.*, (1994). Towards immunity to potato leafroll virus and potato mop top virus by using transgenic and host gene-mediated forms of resistance. *Aspects Applied Biology*, **39**, 189-194.

Baulcombe, D. (1994a). Replicase-mediated resistance: a novel type of virus resistance in transgenic plants. *Trends Microbiol*, **2**, 60-63.

Baulcombe, D. (1994b). Novel strategies for engineering virus resistance in plants. *Current Opinion Biotechnology*, **5**, 117-124.

Baulcombe, D., Bendahmane, A., Kanyuka, K, *et al.*, (1994). Resistance to PVX in potato. Seventh International Symposium on Molecular Plant-Microbe Interactions, Edinburgh. Abstracts S37, p12.

Beachy, R.N., Loesch-Fries, S. & Tumer, N.E. (1990). Coat protein-mediated resistance against virus infection. *Annual Review Phytopathology*, **28**, 451-474.

Becker, F., Buschfeld, E., Schell, J. *et al.*, (1993). Altered response to viral infection by tobacco plants perturbed in ubiquitin system. *Plant Journal*, **3**, 875-881.

Berger, P.H., Hunt, A.G., Domier, L.L. *et al.*, (1989) Expression in transgenic plants of a viral gene product that mediates insect transmission of potyviruses. *Proceedings of the National Academy of Sciences of the USA*, **86**, 8402-8406.

Blanc, S., Schmidt, I., Kuhl, G. *et al.*, (1993). Paracrystalline structure of cauliflower mosaic virus aphid transmission factor produced both in plants and in a heterologous system and relationship with a solubilized active form. *Virology*, **197**, 283-292.

Bol, J.F., Linthorst, H.J.M. & Cornelissen, B.J.C.(1990). Plant pathogenesis-related proteins induced by virus infection. *Annual Review Phytopathology*, **28**, 113-138.

Braun, C.J. & Hemenway, C.L. (1992). Expression of amino-terminal portions or full length viral replicase genes in transgenic plants confers resistance to PVX infection. *Plant Cell*, **4**, 735-744.

Chen, Z.C., White, R.F., Antoniw, J.F. *et al.*, (1991). Effect of pokeweed antiviral protein (PAP) on the infection of plant viruses. *Plant Pathology*, **40**, 612-620.

Chen, Z.C., Antoniw, J.F., Lin, O. *et al.*, (1993). Expression of pokeweed (*Phytolacca americana*) antiviral protein cDNA in *E.coli* and its antiviral activity. *Physiology Molecular Plant Pathology*, **42**, 237-247.

Citovsky, V., Wong, M.L., Shaw, A.L. *et al.*, (1992). Visualization and characterization of TMV movement protein binding to single-stranded nucleic acids. *Plant Cell*, **4**, 397-411

Culver, J.N., Stubbs, G. & Dawson, W.O. (1994). Structure-function relationship between TMV coat protein and hypersensitivity in Nicotiana sylvestris. *Journal of Molecular Biology*, **242**, 130-138.

Dale,P.J. (1995). R & D regulation and field trialling of transgenic crops. Trends in *Bio/Technology*, **13**, 398-403.

Deom, C.M., Lapidot, M. & Beachy, R.N. (1992). Plant virus movement proteins. *Cell*, **69**, 221-224.

Deom, C.M., Schubert, K.R., Wolf, S. *et al.*, (1990). Molecular characterization and biological function of the movement protein of TMV in transgenic plants. *Proceedings of the National Academy of Sciences of the USA*, **87**, 3284-3288.

Deom, C.M., Oliver, M.J. & Beachy,R.N. (1987). The 30-kilodalton gene product of TMV potentiates virus movement. *Science*, **237**, 389-394.

Edington, F., Dixon, R.A. & Nelson,R.S. (1992). Ribozymes: descriptions and uses. In *Transgenic Plants: Fundamentals and Applications* (A. Hiatt Ed.), pp. 301-323. Marcel Dekker Inc: New York.

Falk, B.W. & Bruening, G (1994). Will transgenic crops generate new viruses and new diseases. *Science*, **263**, 1395-1396.

Fritig, B., Kaufmann, S., Dumas, B. *et al.*, (1987). Mechanism of the hypersensitive reaction of plants. In *Plant Resistance to Viruses* (B.D.Harrison, Ed.), pp. 92-103. Ciba Foundation Symposium 133. John Wiley and Sons: Chichester.

Gerlach, W.L., Haseloff, J.P., Young, M.J. *et al.*, (1990). Use of plant virus satellite RNA sequences to control gene expression. In *Viral genes and Plant Pathogenesis* (T.P. Pirone and J.G. Shaw, Eds), pp. 177-184. Springer-Verlag: New York

Golemboski, D.B., Lomonossoff, G.P. & Zaitlin, M. (1990). Plants transformed with a TMV reconstructed gene sequence are resistant to the virus. *Proceedings of the National Academy of Sciences of the USA*, **87**, 6311-6315.

Haan, P., de Gielen, J.J.L., Prins, M. *et al.*, (1992). Characterization of RNA-mediated resistance to tomato spotted wilt virus in transgenic tobacco plants. *Bio/Technology*, **10**, 1133-1137.

Harrison, B.D. (1992). Genetic engineering of virus resistance: a successful genetical alchemy. *Proceedings of the Royal Society of Edinburgh*, **99B**, 61-77.

Harrison, B.D. & Murant, E.A. (1989). Genetic engineering of virus resistance. *Report of the Scottish Crop Research Institute for 1988*, pp. 164-166.

Hartley, M.R., Legname, G., Osborn, R. *et al.*, (1991). Single-chain ribosome inactivating proteins from plants depurinate *E. coli* 23S ribosomal RNA. *FEBS Letters*, **20**, 65-68.

Hayakawa, T., Zhu, Y., Itoh, K. *et al.*, (1992). Genetically engineered rice resistant to rice stripe virus, an insect-transmitted virus. *Proceedings of the National Academy of Sciences of the USA*, **89**, 9865-9869.

Hull, R. (1994). Molecular biology of plant virus-vector interactions. *Advances in Disease Vector Research*, **10**, 361-386.

Hull, R. (1989). The movement of viruses in plants. *Annual Review Phytopathology*, **27**, 213-240.

Jorgensen, R. (1990). Altered gene expression in plants due to *trans* interactions between homologous genes. *Tibtech*, **8**, 340-344.

Kawchuk, L.M., Martin, R.R. & McPherson, J. (1941). Sense and antisense RNA-mediated resistance to potato leafroll virus in Russet Burbank potato plants. *Molecular Plant-Microbe Interactions*, **4**, 247-253.

Kim, J.W., Sun, S.S.M. & German,T.L. (1994). Disease resistance in tobacco and,tomato plants transformed with the tomato spotted wilt virus nucleocapsid gene. *Plant Disease*, **78**, 615-621.

Kollar, A., Dalmay, T. & Burgyan, J. (1993). Defective interfering RNA-mediated resistance against cymbidium ringspot tombusvirus in transgenic plants. *Virology*, **193**, 313-318.

Koonin, E.V., Mushegian, A.R., Ryabov, E.V. *et al.*, (1991). Diverse groups of plant RNA and DNA viruses share related movement proteins that may possess chaperone-like activity. *Journal of Genetic Virology*, **72**, 2895-2903.

Kunik, T., Salomon, R., Zamir, D. *et al.*, (1994). Transgenic tomato plants expressing the tomato yellow leaf curl virus capsid protein are resistant to the virus. *Bio/Technology*, **12**, 500-504.

Kulaeva, O., Fedina, A.B., Burkhanova, E.A. *et al.*, (1992). Biological activities of human interferon and 2-5 oligoadenylates in plants. *Plant Molecular Biology*, **20**,383-393.

Lapidot, M. Gafny, R. Ding, B. *et al.*, (1993). A dysfunctional movement protein of TMV that partially modifies the plasmodesmata and limits virus spread in transgenic plants. *Plant Journal*, **4**, 959-970.

Lecoq, H., Ravelonandro, M., Wipf-Scheibel, C. *et al.*, (1993). Aphid transmission of a non-aphid transmissible strain of zucchini yellow mosaic potyvirus from transgenic plants expressing the capsid protein of plum pox potyvirus. *Molecular Plant-Microbe Interactions*, **6**, 403-406.

Lindbo, J.A. & Dougherty, W.G. (1992a). Pathogen-derived resistance to a potyvirus : immune and resistant phenotypes in transgenic tobacco expressing altered forms of a potyvirus coat protein nucleotide sequence. *Molecular Plant-Microbe Interactions*, **5**, 144-153.

Lindbo, J.A. & Dougherty, W.G. (1992b). Untranslatable transcripts of the tobacco etch virus coat protein gene sequence can interfere with TEV replication in transgenic plants and protoplasts. *Virology*, **189**, 725-733.

Lindbo, J.A., Silva-Rosales, L., Proebsting, W.M. *et al.*, (1993). Induction of a highly specific antiviral state in transgenic plants: implications for regulation of gene expression and viral resistance. *Plant Cell*, **5**, 1749-1759.

Linthorst, H.J.M. (1991). Pathogenesis-related proteins in plants. *Crit Revs Plant Sci*, **10**, 123-150.

Lodge, J.K., Kaniewski, W.K. & Tumer, N.E. (1993). Broad-spectrum virus resistance in transgenic plants expressing pokeweed antiviral protein. *Proceedings of the National Academy of Sciences of the USA*, **90**, 7089-7093.

Longstaff, M., Brigneti, G., Boccard, F. *et al.*, (1993). Extreme resistance to PVX infection in plants expressing a modified component of the putative viral replicase. *EMBO Journal*, **12**, 379-386.

Lucas, W.J. & Gilbertson, R.L. (1994). Plasmodesmata in relation to viral movement within leaf tissues. *Annual Review Phytopathology*, **32**, 387-411.

Maiti, I.B., Murphy, J.F., Shaw, J.G. *et al.*, (1993). Plants that express a potyvirus proteinase gene are resistant to viral infection. *Proceedings of the National Academy of Sciences of the USA*, **90**, 6110-6114.

Malyshenko, S.I., Kondakova, O.A., Nazarova, J.V. *et al.*, (1993). Reduction of TMV accumulation in transgenic plants producing non-functional viral transport proteins. *Journal of Genetic Virology*, **74**, 1149-1156.

Maule, A. (1991). Virus movement in infected plants. *Crit Revs Plant Sci*, **9**, 457-473.

Moffat, A.S. (1994) Mapping the sequence of disease resistance. *Science*, **265**, 1804-1805.

Mueller, E., Gilbert, J., Davenport, G., Brigneti, G., & Baulcombe, D.C. (1995). Homology-dependent resistance: transgenic virus resistance in plants related to homology-dependent gene silencing. *Plant Journal*, **7**, 1001-1013.

Pang, S.Z., Slightom, J.L. & Gonsalves,D. (1993). Different mechanisms protect transgenic tobacco against tomato spotted wilt and impatiens necrotic spot tospoviruses. *Bio/Technology*, **11**, 819-824.

Plafker, T. (1994). First biotech safety rules don't deter Chinese efforts. *Science*, **266**, 966-967.

Poirson, A., Turner, A.P., Giovane, C. *et al.*, (1993). Effect of the alfalfa virus movement protein expressed in transgenic plants on the permeability of plasmodesmata. *Journal of Genetic Virology*, **74**. 2459-2461.

Powell-Abel, P., Nelson, R.S., Be, D., *et al.*, (1986). Delay of disease development in transgenic plants that express the TMV coat protein gene. *Science*, **232**, 738-743.

Shah, D., Lawson, C. Kaniewski, W. *et al.*, (1994). Russet Burbank genetically improved for resistance to potato leafroll virus and late blight. Seventh International Symposium on Molecular Plant-Microbe Interactions, Edinburgh. Abstracts, S48, p 14.

Shah, D. M., Rommens, C. T. M. & Beachy, R. N. (1995). Resistance to diseases and insects in transgenic plants; progress and applications to agriculture. *Trends in Biotechnology*, **13**, 362-368.

Silva-Rosales, L., Lindbo, J.A. & Dougherty, W.G. (1994). Analysis of transgenic tobacco plants expressing a truncated form of a potyvirus coat protein nucleotide sequence. *Plant Molecular Biology*, **24**, 929-939.

Stanley, J., Frischmuth, T. & Ellwood, S. (1990). Defective viral DNA ameliorates symptoms of geminivirus infection in transgenic plants. *Proceedings of the National Academy of Sciences of the USA*, **87**, 6291-6295.

Stirpe, F., Barbieri, L., Batelli, M.G. *et al.*, (1992). Ribosome-inactivating proteins from plants: present status and future prospects. *Bio/Technology*, **10**, 405-412.

Tavladoraki, P., Benvenuto, E., Trinca, S. *et al.*, (1993). Transgenic plants expressing a functional single chain Fv antibody are specifically protected from virus attack. *Nature*, **366**, 469-472.

Taylor, S., Massiah, A., Lomonossoff, G. *et al.*, (1994). Correlation between the activities of five ribosomal-inactivating proteins in depurination of tobacco ribosomes and inhibition of TMV infection. *Plant Journal*, **5**, 827-835.

Tien, P., Zhang, X., Qiu, B. *et al.*, (1987). Satellite RNA for the control of plant diseases caused by CMV. *Annals of Applied Biology*, **111**, 143-152.

Truve, E., Aaspollu, A., Honkanen, J. *et al.*, (1993). Transgenic potato plants expressing mammalian 2-5 oligoadenylate synthetase are protected from PVX infection under field conditions. *Bio/Technology*, **11**, 1048-1052.

van den Elzen, P.J.M., Jongedijk, E., Melchers, L.S. *et al.*, (1993). Virus and fungal resistance: from laboratory to field. *Philosophical Transactions of the Royal Society. Series B*, **342**, 271-278.

van Loon, L.C. (1985). Pathogenesis-related proteins. *Plant Molecular Biology*, **4**, 111-116.

Ward, E.R., Uknes, S.J., Williams, S.C. *et al.*, (1991). Coordinated gene activity in response to agents that induce systemic acquired resistance. *Plant Cell*, **3**, 1085-1094.

Wilson, T.M.A. (1993). Strategies to protect crop plants against viruses: pathogen-derived resistance blossoms. *Proceedings of the National Academy of Sciences of the USA*, **90**, 3134-3141.

Wuethrich, B. (1994). Will altered plants breed deadlier diseases? *New Scientist*, **142** (No 1919), 15.

5. MODIFYING RESISTANCE TO PLANT-PARASITIC NEMATODES

P.R. Burrows
(IACR-Rothamsted)

5.1 BACKGROUND

Plant parasitic nematodes are a diverse group of microscopic organisms that feed on the living cells of their host plants. Most species are migratory browsers that move over the surface or through the tissues of their host plants and kill the cells on which they feed before moving on. However, some nematode groups have adopted a sedentary life style and instead of browsing on cells, they induce and maintain specialised feeding sites within their hosts (Jones, 1981).

Many species of plant parasitic nematodes pose a significant threat to crop production worldwide. The financial loss incurred in world agriculture due to nematodes is difficult to determine accurately but is thought to be $5.8 billion per year in the United States alone.

5.1.1 Economically Important Nematodes in European Agriculture.

5.1.1.1 Cyst nematodes; *Globodera* spp and *Heterodera* spp.

In northern Europe, the potato-cyst nematodes (PCN) *Globodera rostochiensis* and *G. pallida* are the most damaging pests of potatoes. Both species occur in most (if not all) ware and seed potato growing areas and, based on European potato production data, are estimated to cause yield losses valued at £200-300 million per year. The cost of nematicides to control these nematodes and the enforcement of the statutory controls has not been properly considered but is in excess of £100 million per year.

The beet cyst nematode *Heterodera schachtii* is recognised as a severe limitation to the growth of sugar-beet and it was responsible for the near collapse of the European sugar-beet industry in the late 19th century. Today, most beet growing areas are infested and control relies mainly on nematicides and long rotations between susceptible crops. Other species of cyst nematodes such as *H. trifolii* (clover cyst nematode), *H. avanae* (cereal cyst nematode), *H. cruciferae*, (brassica cyst nematode) and *H. goettingiana* (pea cyst nematode) cause important local problems.

5.1.1.2 Root-knot Nematodes; *Meloidogyne* spp.

Root-knot nematodes are primarily a problem in the warmer mediterranean countries where they parasitise crops such as peppers, tomatoes and grapes. However, *Meloidogyne chitwoodi* is rapidly becoming a significant pest on potatoes in the Netherlands. Economic loss due to root-knot nematodes has not be estimated for many regions, and the figures that are available are considered to be underestimates as many root-knot problems go unidentified or misclassified in the poorer southern countries.

5.1.1.3   Root-lesion nematodes; *Pratylenchus* spp.

Root-lesion nematodes occur widely in European soils where they cause local economic damage to a wide range of crops, ornamentals, soft and top fruits.   No economic treatment is available on the field scale.

5.1.1.4   Virus vectors; *Xiphinema, Longidorus, Trichodorus* and *Paratrichodorus.*

The genera *Xiphinema, Longidorus, Trichodorus* and *Paratrichodorus* all contain species that are known vectors of damaging plant-pathogenic viruses. These nematodes transmit viruses mainly to woody and herbaceous plants such as strawberry, blackcurrent, raspberry, plum and cherry. Three of the above genera, *Longidorus, Trichodorus* and *Paratrichodorus*, are also responsible for docking disorder of sugar-beet, although this disease is primarily due to high nematode burdens rather than virus transmission.

5.1.2   Nematode Control

The control of plant parasitic nematodes relies heavily on the use of toxic and expensive chemical control agents.   In recent years concern about the presence of certain nematicides or their breakdown products in soil or ground water has led to the banning of several chemicals which were once in common use (Thomason, 1987). Pressure from the powerful environmental lobby is an important factor in the current unpopularity of the chemical control of nematodes. The use of nematode resistant crops in rotation schemes is an effective means of controlling nematodes without toxic chemicals, but despite much effort very few commercially viable cultivars have come out of breeding programmes.   There are still many important crops or wild germplasm in which nematode resistance has not been found or has been difficult to develop.

5.2   ENGINEERING RESISTANCE TO NEMATODES.

Recent advances in plant biotechnology now provide the means by which better and much wider use could be made of resistant crop cultivars to control nematodes by engineering novel forms of nematode resistance.

5.2.1   Engineering Resistance to Sedentary Nematodes

Most research concerning engineered nematode resistance has concentrated on the sedentary nematodes that need to induce and maintain feeding sites within their host plants. These groups, especially *Globodera, Heterodera* and *Meloidogyne*, have received most attention because not only are they globally the most economically damaging plant parasitic nematodes, but also their total dependence on a specialised feeding site provides a convenient target for disruption.

In order to induce and maintain their feeding sites these highly advanced parasites modify host gene expression, either activating or repressing a wide range of host genes within the few root cells that make up the feeding sites. The basis of most transgenic approaches to achieving resistance depends on the identification and cloning of plant gene promoters capable of directing the expression of transgenes within the feeding sites, with little or no expression elsewhere in the plant. A few such promoters have been reported (Gurr *et al.*, 1992 : Opperman *et al.*, 1994). Perhaps the most interesting is that derived from a tobacco gene (Tob RB7) that is selectively activated by *Meloidogyne* nematodes, and which probably codes for a protein, possibly an aquaporin (Chrispeels & Maurel, 1994), involved in the structure of a water channel through the plant cell membranes. The specificity for its expression in nematode-infected cells was remarkably increased when it was truncated from 1.8 k base pairs (bp) to a mere 300 bp fragment from the 3' end (Taylor *et al.*, 1992; Opperman *et al.*, 1994). The importance of such feeding-site specific promoters cannot be overstated, as many promoters commonly used in plant biotechnology, e.g. the CaMV 35S from cauliflower mosaic virus, are repressed in the feeding sites and are thus ineffective.

Feeding-site specific promoters are currently being used in the following experimental approaches to resistance:

5.2.1.1   Antisense and gene co-suppression.

The feeding site promoter is used to activate the expression of an antisense construct capable of disrupting the function of an important gene involved in essential cell metabolism or in an essential physiological function. Infection then leads to the disruption of feeding sites with minimal effects on other parts of the plant. Although no working examples of this approach have yet been published, limited success has been achieved with *Meloidogyne* using an antisense construct to the water channel gene Tob RB 7 (C.H. Opperman. personal communication). Work using other genes is currently being conducted at IACR-Rothamsted.

5.2.1.2   Cytotoxic genes.

The specific expression of cytotoxic genes in the feeding sites would also be expected to destroy these sites specifically. The toxic RNAase barnase, derived from *Bacillus amyloliquefaciens* (see Mariani *et al.*, 1992) is often considered in this respect. This approach requires very 'tight' promoters, as toxin expression elsewhere in the plant will be damaging or even lethal to the plant. Barnase constructs, under the control of the modified promoter derived from the water channel gene Tob RB7, resulted in the first recorded example of engineered resistance to nematodes (C.H. Opperman, personal communication).

A variation on this theme, which circumvents the need for an extremely specific promoter, exploits the repression of the CaMV 35S promoter when present in feeding-site cells (Sijmons, 1993). In this system, the so called "two component system", the feeding site promoter activates synthesis of a cytotoxin while the CaMV 35S promoter is used for the constitutive expression of an inhibitor of the cytotoxin. In theory, any leakage of expression of the toxin in plant tissues other than the feeding sites is nullified by the inhibitor. The nematode's feeding sites are thus the only cells in which the toxin is expressed without the inhibitor. This approach is advocated by Sijmons (1993) and his colleagues; they have constructed and are testing *Arabidopsis* plants in which the cytotoxin is barnase, and its constitutively expressed neutralising agent is the RNAase-inhibitor barstar (Hartley, 1988).

40

5.2.1.3   Plant produced antibodies (Plantibodies)

Sedentary nematodes are thought to initiate and maintain their feeding site by the injection of salivary proteins into the feeding site cells (Jones, 1981). Antibodies raised to one or more of these proteins are being cloned into plants, as antibody fragments, under the control of feeding site promoters in an attempt to neutralise or to reduce the functional concentration of the nematode proteins. As a result of this it is expected that feeding sites will not be initiated or will degenerate.   (F. Gommers, personal communication).

Alternative approaches are that antibodies to specific nematode sensory organs, or even to plant proteins that may be necessary for the induction of feeding sites, be introduced into plants under the control of specific promoters (see Burrows & Jones, 1993). These approaches however, have been considered less promising and have received little attention.

5.2.2   Transgenic Approaches to Engineering Resistance to Sedentary and Browsing Nematodes.

The following examples of engineered resistance are equally applicable to sedentary and browsing nematodes. However, it should be realised that, while general constitutive or root specific promoters will be adequate for expression of anti-nematode transgenes to combat the browsing nematodes, the sedentary groups will still require promoters specific for feeding site cells for maximal effect.

5.2.2.1   Enzyme inhibitor and Lectin genes.

Two main classes of plant-derived proteins, enzyme inhibitors and lectins, are being exploited to produce plants resistant to insects (Boulter *et al.*, 1990). Genes encoding both classes of molecules have been isolated and expressed in transgenic plants with some success (see section 7.3 - 7.5).

It is not possible to do straightforward feeding and toxicity trials with plant parasitic nematodes as it is with herbivorous insects, and consequently it is more difficult to choose for instance, proteinase inhibitors which are potentially good candidates for introducing into appropriate plants. Urwin *et al.* (1995) used *Caenorhabditis elegans* as a model nematode to test native and engineered cystine-proteinase inhibitors. Based on apparent toxicity towards *C. elegans*, native and engineered oryzacystatin were expressed in tomato hairy roots and shown to confer resistance to the potato cyst nematode *G. pallida*. Another productive strategy, at least in the short term, is to screen anti-insect transgenic plants, as they become available, for resistance against plant parasitic nematodes. Any genes found to confer resistance could, if necessary, be modified to change the level and location of expression to target specific nematode groups.

Certain plant lectins, such as concanavalin A, have been shown to bind to the sensory apparatus (amphids) of plant parasitic nematodes and so disrupt host location. They may also be implicated in binding to gut wall components and interfering with gut function. Preliminary studies at IACR-Rothamsted and the Scottish Crop Research Institute indicate that the snowdrop lectin, GNA, is effective against plant nematodes in transgenic plants. The mode of action is not yet understood.

Clearly, the identification and use of inhibitor and lectin genes could be valuable for conferring nematode resistance. This approach has a number of advantages; the gene products are essentially non-toxic to humans (indeed some are derived from food plants), the resistance is typically broad spectrum, and using a mulitmechanistic/multigene approach it is likely to be more durable than single gene resistance.

5.2.2.2   *Bacillus thuringiensis* (Bt) toxins.

Transgenic plants which express genes for Bt toxins are resistant to specific insects (see section 7.2), but the narrow specificity of individual Bt toxins probably precludes any cross protection against nematodes. It is unclear however, if this has been tested for extensively with existing transgenic plants. Toxins from several strains of *B. thuringiensis* have been shown in laboratory tests to exhibit various degrees of toxicity towards some free living or parasitic nematodes (see Burrows and Jones, 1993). In one case (Wharton and Bone, 1981), *in vitro* exposure to a Bt toxin caused aberrations in the lipid layers and outer membranes of nematode eggs and embryos, suggesting a contact rather than a digestion-dependant toxicity. More recently, nematode active toxins have been claimed in patents (e.g. Schrepf *et al.*, 1992; see Koziel *et al.*, 1993), although unpublished, anecdotal evidence questions the reliability of some of these claims. Attempts have been made to introduce some of these toxin genes into plants. So far, to our knowledge, the success obtained with insect resistance has not been repeated for parasitic nematodes.

Nematicidal activity has been claimed for an exotoxin produced by a strain of *B. thuringiensis*, but it was too small to be useful in soil treatments (Devidas and Rehberger, 1992). These exotoxins are nucleotide analogues which interfere with the normal functioning of DNA-dependent RNA polymerases. Unlike the endotoxins, they are heat stable, and their toxicity is unspecific: their genes are carried on bacteria plasmids which have been removed from most formulations of Bt toxins which are used commercially.

Some species of predatory fungi produce toxins effective in paralysing and killing nematodes (e.g. Barron and Thorne, 1987). Their structure is unknown. They are most likely to be fungal secondary metabolites produced by a sequence of enzymes; but should they prove to be primary gene products that are specifically toxic to nematodes, they will be obvious candidates for genetic manipulation.

5.2.3   Mapping and Isolating Natural Nematode Resistance Genes.

Progress in genetic mapping may soon lead to the isolation of natural nematode resistance genes that could be transferred between crops. Particular interest has been shown in the chromosomal location of resistance genes to *G. rostochiensis* (Barone *et al.*, 1990), to *H. schachtii* in beet (Jung *et al.*, 1992) and to *Meloidogyne* spp. in tomatoes (the Mi gene) (Messeguer *et al.*, 1991). The Mi gene that confers resistance to *Meloidogyne* spp. is likely to have been isolated and tested in transgenic systems within two years.

5.3   CONCLUSIONS

The first plants with engineered nematode resistance have been produced and are being tested in Europe and the US, and it is felt that, like virus resistance, there is a realistic opportunity of creating useful and durable resistance in a number of crops.

The greatest nematode problems facing northern European agriculture are caused by sedentary endoparasites, principally the potato cyst nematodes and to a lesser extent the beet cyst nematode. Work should be supported that aims to develop specific plant promoters that facilitate high levels of expression of transgenes in the feeding sites of these pests. Research programmes using such promoters in association with cytotoxic genes, antisense constructs and enzyme inhibitors are likely to be the most productive. Resistance to cyst nematodes would be of most use to potato and sugar-beet breeders.

It is important that research is supported that aims to develop anti-nematode gene constructs containing genes that are toxic or deleterious to browsing nematodes. The programmes investigating the use of enzyme inhibitors and lectins are currently the most realistic and should be particularly encouraged.

5.4   REFERENCES

Barone, A., Ritter, E., Schnachtschabel, U., Debener, T., Salamini, F. & Gebhardt, C. (1990). Localisation by restriction fragment length polymorphism mapping of a major dominant gene conferring resistance to the potato cyst-nematode *Globodera rostochiensis*. *Molecular and General Genetics*, **224**, 177-182.

Barron, G.L. & Thorne, R.G. (1987). Destruction of nematodes by species of *Pleurotus*. *Canadian Journal of Botany*, **65**, 774-778

Boulter, D., Gatehouse, J.A., Gatehouse, A.M.R. & Hilder, V.A. (1990).   Genetic engineering of plants for insect control. *Endeavour*, **14**, 185-190.

Burrows, P.R. & Jones, M.G.K. (1993). Cellular and Molecular approaches to the control of plant parasitic nematodes. In: *Plant Parasitic Nematodes in Temperate Agriculture*, Eds. K. Evans, D.L. Trudgill and J.M. Webster. pp. 648. CAB International, Wallingford, Oxon, UK. University Press Cambridge 1993. ISBN 0 58198 808 3

Chrispeels, M.J. & Maurel, C. (1994). Aquaporins: the molecular basis of facilitated water movement through living plant cells. *Plant Physiology*, **105**, 9-13.

Devidas, P. & Rehberger, L.A. (1992). The effects of exotoxin (Thuringiensin) from *Bacillus thuringiensis* on *Meloidogyne incognita* and *Caenorhabditis elegans*. *Plant and Soil*, **145**, 115-120

Gurr, S.J., McPherson, M.J., Atkinson, H.J & Bowles, D.J. (1992).   Plant parasitic nematode control. International Patent Application WO 92/04453.

Hartley, R.W. (1988). Barnase & barstar: expression of its cloned inhibitor permits expression of a cloned ribonuclease. *Journal of Molecular Biology*, **202**, 913-915

Jones, F.G.W. & Jones, M.G. (1984). *Pests of field crops*. Third edition. Edward Arnold, London. ISBN 0 7131 2881 X

Jones, M.G.K. (1981). Host cell responses to endoparasitic nematode attack: structure and function of giant cells and syncytia. *Annals of Applied Biology*, **97**, 353-372.

Jung, C., Claussen, U., Horsthemke, B., Fischer, F. & Hermann, R.G. (1992). A DNA library from an individual *Beta patellaris* chromosome conferring nematode resistance obtained by microdissection of meiotic metaphase chromosomes. *Plant Molecular Biology*, **20**, 503-511

Koziel, M.G., Carozzi N.B., Currier, T.C., Warren, G.W. & Evola S.V. (1993). The insecticidal crystal proteins of *Bacillus thuringiensis*: past, present and future uses. *Biotechnology and Genetic Engineering Reviews*, **11**, 171-228

Mariani, C., Gossele,V., De Beuckeleer, M., De Block, M., Goldberg, R.B., De Greef, W. & Leemans, J. (1992). A chimaeric ribonuclease-inhibitor gene restores fertility to male sterile plants. *Nature*, **357**, 384-387

Messeguer, R., Ganal, M., de Vicente, M.C., Young, N.D. & Tinkle, S.D. (1989). High resolution RFLP map around the root-knot nematode resistance gene (Mi) in tomato. *Theoretical and Applied Genetics*, **82**, 529-536

Opperman, C.H., Taylor, C.G. & Conkling, M.A. (1994). Root-knot directed expression of a plant root specific gene. *Science*, **263**,221-223.

Roberts, P.A. (1992). Current status of the availability, development and use of host plant resistance to nematodes. *Journal of Nematology*, **24**, 213-227

Sijmons, P.C. (1993). Plant-nematode interactions. *Plant Molecular Biology*, **23**, 917-931.

Taylor, C.G., Song, W., Opperman, C.J. & Conkling, M.A. (1992). Characterization of a nematode-responsive plant gene promoter. *Journal of Nematology*, **24**, 621.

Thomason, I.J. (1987). Challenges facing nematology: Environmental risks with nematicides and the need for new approaches. In: *Vistas on Nematology*. Eds. J.A. Veech and D.W. Dickson. Hyattsville, Maryland. pp 469-476

Wharton, D.A. & Bone, L.W.(1989). *Bacillus thuringiensis* israelensis toxin affects egg shell ultrastructure of *Trichostrongylus colubriformis*. *Invertebrate Reproduction and Development*, **15**, 155-158

6.   RESISTANCE TO MOLLUSCS

I.F. Henderson and D.M. Glen
(IACR-Rothamsted and IACR-Long Ashton)

6.1   BACKGROUND

The main crops damaged by terrestrial molluscs in temperate climates are cereals, potatoes oilseed rape, sunflower and field vegetables. Worldwide, the list is longer, and citrus crops and rice are badly affected in Mediterranean and tropical regions. It is difficult to obtain estimates of the cost of the damage. NW Europe has the largest market for slug controlling chemicals and the UK the largest share of this market. This regional usage, however, may reflect more on the cost and efficacy of the control treatments rather than on the extent and importance of the damage done by slugs. In the UK, some 290,000 hectares of crops were treated with approximately 100 tonnes of molluscicide in 1992 (British Agrochemicals Association figures). Of these 100 tonnes, cereals received 57%, potatoes 28%, and oilseed rape about 16%. These usages are small compared with those of other pesticides, and, in monetary terms, cost about £8 million compared to about £50 million for insecticides and about £132 million for fungicides. Moreover the year-to-year usage is very variable and was, for instance, three times larger in 1988 than in 1990. The importance of slugs as a problem is perhaps better illustrated in the expression of farmers' opinions rather than in molluscicide use: the views of cereal farmers belonging to the Long Ashton Members Association were surveyed in 1986-7, when 46% of them identified slugs as the major perceived pest in wheat crops and 24% identified slugs as a major pest, second only to aphids, in barley (Glen, 1989). The surveys were admittedly small and localized, but they, as well as surveys of molluscicide usage, suggested that there had been an increase in the importance of slug damage to UK cereals since estimates were made in the 1960s and 70s. One of the reasons for such an increase may be the increased cultivation of oilseed rape, which provides ideal conditions for slug populations to increase: winter wheat, sown straight after an oilseed rape crop is particularly vulnerable (see Glen, 1989). Changes in the harvesting and marketing of potatoes may also have contributed to an increase in the importance of slug-damage in this crop. Beer (1989) describes briefly but graphically how comparatively low levels of damage may be dramatically reflected in a fall in the value of a crop. Fewer than 5% slug damaged potatoes in an otherwise healthy crop, render the crop unsuitable for prepacked sale in supermarkets. Damage, moreover, tends to be more extensive in "white" potatoes such as Maris Piper, which have a high customer appeal.

6.2   CURRENT CONTROL METHODS

Apart from cultural control measures, the best available method of controlling slug damage remains the delivery of stomach-acting molluscicides in edible bait formulations. This is inefficient and creates a risk to non-target organisms, and is under increasing regulatory scrutiny. This consideration, as well as the high cost of registering new molluscicides and the comparative small size of the market, results in relatively little work being devoted to developing new molluscicides. Virtually all the new ones introduced over the last 50 years were originally developed as acaricides or insecticides, so that their development costs were largely underwritten by larger markets. This is true of the carbamate, methiocarb, which together with the unrelated compound metaldehyde has wide commercial use (see Airey *et al.*, 1989). Some metal chelate compounds, developed and patented at IACR-Rothamsted in 1986, were specifically designed for slug toxicity, but so far they have not been developed commercially. IACR-Long Ashton has had more success by developing a biological form of control in the form of a rhabditid nematode which is parasitic upon such common slug species as *Deroceras reticulatum* (Wilson *et al.*, 1993). This organism, an indigenous UK species, was isolated from a colony of stressed slugs, and is able to penetrate into the mantle of its host and kill it, possibly by introducing bacteria which are pathogenic in the presence of the nematode (Glen *et al.*, 1994; Wilson *et al.*, 1994). This method of control requires no licensing as it utilises an indigenous species, and its commercial exploitation has been undertaken by MicroBio Ltd (a wholly owned subsidiary of the Agricultural Genetics Company Ltd which funded the research). The product is distributed and sold on the home garden market by Defenders Ltd under the trade name "Nemaslug" and by Zeneca plc in their "Nature's Friends" range of biological control agents. Limited production capacity confines it at present to the home garden market. It will be necessary to scale up production, reduce costs and improve shelf life of the product before it can be used by arable farmers. Research to establish the principles for effective use of the nematode in arable crops is the subject of current research.

6.3   FUTURE CONTROL METHODS

The search for chemical molluscicides has continued in a number of directions (Airey *et al.*, 1989). Natural toxicants examined include the saponins of African plants which are active against species of water snails (Hostettmann & Marston, 1987). Attempts have been made to exploit the obvious biochemical features such as calcium metabolism and mucus production, that distinguish slugs and snails from other animals. Attempts have also been made to identify any chemical signals, semiochemicals, produced by either mollusc or plant, that modify the animals' behaviour. Perhaps the most promising approach in this direction (Airey *et al.*, 1989) involves a search for antifeedants and repellants produced in plants. Such compounds have often been postulated to explain the selectivity of slug feeding. Johnston & Pearce (1994), for example, related slug-resistance in potatoes to the presence of small amounts of oxidizable phenols and large quantities of the enzyme phenolase in the layer just beneath the skin of potato tubers. When a slug first penetrates the skin, the enzyme acts on the phenols to produce quinones. It is thought that, when ingested by the slug, the quinones inhibit feeding and reduce growth rate. Johnston & Pearce (1994) point out that although quinones have the undesirable effect of causing blackening of the skin of potato flesh when it is cut, blackening is not necessarily associated with this mechanism of resistance to slug damage: only small quantities of phenolic compounds are needed to give resistance, provided that high levels of the phenolase are present to oxidise them rapidly to quinones. Thus it may be possible to use genetic engineering to increase the degree of resistance in susceptible varieties such as Maris Piper by enhancing the level of phenolase. Airey *et al.*, (1989) have screened extracts of over 60 plant species as well as many known arthropod antifeedants, for the ability to make wheat seeds unpalatable to slugs. A number of promising secondary metabolites were detected and examined, but

46

found to have stabilities and volatilities unsuited for use in applied formulations. The bicyclic terpenoid ketone, (+)fenchone, which is related to the more common monoterpene geraniol, was one such compound that is still under consideration. Researchers at CSL have recently identified cinnamamide as a potential antifeedant for slugs (New Scientist 26 November 1994 p23). This compound is a synthetic derivative of cinnamic acid which protects buds of certain varieties of pear tree from bullfinch damage.

The search for low molecular weight, slug-repelling compounds continues at IACR. If they can be identified and shown to be harmless both to other animals and human consumers, they may be able to complement or even replace molluscicide formulations. An alternative and more attractive way of using them would be to manipulate the synthetic capacities of crop plants so as to produce them. However, introducing multigene metabolic pathways into plants demands, among other things, a knowledge of plant metabolism that is lacking in many respects, and it is still a very long term aim. A more direct way of using the techniques of genetic manipulation to produce slug-resistant plants, is to create transgenic plants containing protein inhibitors of molluscan digestive enzymes. It is unclear whether the transgenic plants containing protease inhibitors that are resistant to some insects (see section 7.3), are also resistant to slugs, but it should not be discouraging if they are susceptible. A rational choice of effective digestion inhibitors can only be made when the nature of the enzymes involved in molluscan digestion, and also the conditions in which they operate are known. This necessary background work is being undertaken at IACR-Long Ashton.

6.4 CONCLUSIONS

Molluscs, especially slugs, are important pests of cereals, potato and oilseed crops in north western Europe. Changes in farm practice and also in marketing have increased the extent and cost of the damage that they do. Current control methods using toxic baits are unspecific and becoming less acceptable. A developed method of biological control using nematodes is at present useful only on a small scale, although current research aims to establish the principles for effective economic use in arable crops.
Work on the biology of slugs, their feeding preferences and on the antifeedants present in many plants, may in the long term, suggest safer and more acceptable methods of control than current chemical methods. One promising approach to the development of slug-resistant crop plants is to develop transgenic plants containing inhibitors of molluscan digestive enzymes. This approach needs a better understanding of these digestive enzymes, especially proteases, and the conditions in which they operate.

6.5 REFERENCES

Airey, W.J., Henderson, I.F., Pickett, J.A., Scott, G.C., Stephenson, J.W. & Woodcock, C.M. (1989). Novel chemical approaches to mollusc control. In *Slugs and Snails in World Agriculture*. (Henderson, I.F. Ed.), pp.301-307. BCPC Monograph No.41. British Crop Protection Council, Thornton Heath.

Beer, G.J. (1989). Levels and economics of slug damage in potato crops 1987 and 1988. In *Slugs and Snails in World Agriculture*. (Henderson, I.F. Ed.), pp.101-105. BCPC Monograph, No. 41. British Crop Protection Council, Thornton Heath.

Glen, D.M. (1989). Understanding and predicting slug problems in cereals. In *Slugs and Snails in World Agriculture*. (Henderson, I.F. Ed.), pp.253-262. BCPC Monograph No.41. British Crop Protection Council, Thornton Heath.

Glen, D.M., Wilson, M.J., Pearce, J.D. & Rodgers, P.B. (1994). Discovery and investigation of a novel nematode parasite for biological control of slugs. *Proceedings of the Brighton Crop Protection Conference - Pests and Diseases, 1994*, **2**, 617-624.

Hostettmann, K. & Marston, A. (1987). Antifungal, molluscicidal and cytotoxic compounds from plants used in traditional medicine. *Annual Proceedings of the Phytochemical Society of Europe*, pp. 65-83.

Johnston, K.A. & Pearce, R.S. (1994). Biochemical and bioassay analysis of resistance of potato (*Solanum tuberosum* L.) cultivars to attack by the slug *Deroceras reticulatum* (Muller). *Annals of Applied Biology*, **124**, 109-131.

Wilson, M.J., Glen, D.M. & George, S.K. (1993). The Rhabditid nematode *Phasmarhabditis hermaphrodita* as a potential biological control agent for slugs. *Biocontrol Science and Technology*, **3**, 503-511.

Wilson, M.J., Glen, D.M., Pearce, J.D. & Rogers, P.B. (1994). Effects of different bacterial isolates on growth and pathogenicity of the slug-parasitic nematode *Phasmarhabditis hermaphrodita*. *Proceedings of the Brighton Crop Protection Conference - Pests and Diseases, 1994*, **3**, 1055-1060.

7. MODIFYING RESISTANCE TO INSECT PESTS

W.S. Pierpoint and K.J.D. Hughes
(IACR-Rothamsted)

7.1   BACKGROUND

About 1000 tonnes of insecticide, costing some £50 million, are annually applied to crops grown in the UK: many times more must be used throughout Northern Europe. Although this undoubtedly contributes to high yields and to the apparent quality of the produce, it results in concerns relating to the environment and public health. Genetically manipulating crops so as to enhance or confer an innate resistance will help in reducing pesticide use, and indeed this is a principal reason for advocating it. However, such a programme is not without its own difficulties. Thus the general public is distrustful of genetically engineered products, and those involving insect-toxicants may be perceived as presenting a bigger threat to vertebrate animals and humans than those concerned with the control of microorganisms. Moreover, pest resistance may build up as quickly to genetically engineered toxins as to synthetic insecticides. These disadvantages have been considered carefully together with the advantages by Boulter (1993). His conclusions, like those of other reviewers, are optimistic, but he warns that the successful exploitation of the new technology may require basic changes in attitudes among governments, farmers, marketing organizations and the general public.

Most research work in this area has concerned the transgenic introduction of insecticidal toxins, usually proteins, that are derived either from bacteria or from other plants. It is this work which has produced all the commercial and near-commercial crop varieties. More long term work aimed at circumventing some of the disadvantages of this approach, involves manipulating the secondary metabolism of plants to produce toxic metabolites where and when they are needed. An even more radical long term approach aims to elucidate the intricate processes of the chemical signalling that occur between a plant and its predators, and seeks to manipulate secondary metabolism to modify these signals so as to disrupt insect behaviour. The success of this approach will almost certainly require the changes in attitude that Boulter (1993) anticipated.

7.2   RESISTANCE CONFERRED BY TOXIC PROTEINS

7.2.1   Resistance Conferred by *Bacillus thuringiensis* Toxins.

The insect toxins produced in the spores of the soil bacterium, *Bacillus thuringiensis* (Bt), are obvious and almost ideal candidates for inserting into crop plants to confer resistance to insect pests (Gasser & Fraley, 1989). They are single, unmodified proteins coded for by single genes, and over 50 have been characterised (Kolziel *et al.,* 1993; Peferoen, 1992). Their toxicity is high and usually specific to particular orders of insects, for example Lepidoptera, Diptera, Coleoptera, depending on the bacterial strain from which they were derived. Toxicity depends upon proteolysis in the insect gut to give an active fragment, and on the presence of specific receptors in the epithelial cells lining the midgut. The toxins have no, or very few (Goldberg & Tjaden; 1990), adverse effects on mammals or birds. Moreover, as natural bio-degradable substances they have been accepted and used in commercial insecticidal sprays for over 30 years without producing any undesirable ecological effects. It is not surprising that they have attracted the major effort of academic and industrial scientists in the design of insect-resistant plants.

Genes coding for different Bt toxins have been introduced along with constitutive promoters into a range of crop plants including tobacco, tomato, cotton, maize, rice and potatoes, and into apple, walnut and poplar saplings, as well as into chrysanthemum calluses. In all cases, resistance against some relevant insect pest has been demonstrated, sometimes to a degree that has warranted field trials. A major initial difficulty was in getting the Bt genes expressed at adequate levels. This has been accomplished very successfully by a number of strategies which include selecting the most effective promoter, truncating the gene so that it produces only the active protein subunits (M_r about 60000), replacing DNA sequences which were found to confer instability on the corresponding mRNAs and modifying the genes so that they contained the nucleotide codons rich in C+G that are preferred by plants. The most successful genes are thus synthetic, highly refined DNA structures which have low sequence homology (e.g. 65%) with the native bacterial genes (Fischoff et al., 1987; Adang et al., 1993; Koziel et al., 1993). Plants have been transformed with these genes to produce up to 0.4% (Koziel et al., 1993) of their extractable, soluble protein as toxin. The level of insect resistance conferred on cotton crops can be the equivalent of a weekly treatment of the crop with a toxin spray formulation. Published results as well as those of unpublished field trials (Fraley; 1992) suggest that the ability to produce toxin is genetically stable and inherited as a dominant Mendelian trait.

By 1990, it was thought that at least seven US companies had conducted field trials on the first generation of Bt toxin-producing transgenic plants (Goldberg & Tjaden, 1990). The results are not all published, presumably for commercial reasons, but those that are give generally encouraging, although not ideal, results (Koziel et al., 1993). Results of field tests of plants with higher levels of toxin production are also incompletely published, but they are sufficient to encourage reviewers (Koziel et al., 1993) to predict that commercial crops will be available in the near future, and more specifically that "within the next 2-5 years, the first commercial crops are expected to be cotton and maize" (see also Fraley, 1992). There is, however, some concern and debate as to how these crop plants should be used to the best advantage. Resistance to Bt toxins has developed in some target insects, albeit very slowly, in many parts of the world where Bt sprays have been used intensively (Gibbons, 1991; Koziel et al., 1993; Tabashnik, 1994). In the Philippines and in Hawaii, populations of the diamondback moth (Plutella xylostella), which is a pest of watercress and crucifer crops, has developed a 20-fold resistance to the toxin sprays. Much higher degrees of resistance have been produced in the laboratory under more intense conditions of selection both with this moth and with other pests including the Indianmeal moth (Plodia interpunctella), a pest of stored grain. Both field resistance and laboratory-induced resistance are thought to be due to similar mechanism, a change in the receptor sites that decreases their affinity for toxin (Ferre et al., 1991). The field resistance that developed in the diamondback moth was to only one of the forms of the toxin present in the spray. However, experience with other insect species suggests that developing resistances could be more complex, and will involve

50

other mechanisms and also cross resistances between different Bt toxins (McGaughey & Whalon, 1992; Tabashnik, 1994). Moreover, Bt resistance, as in the diamondback moth, is relatively stable, and although mostly genetically recessive, is not quickly lost in the absence of exposure to toxin.

Bt toxins are regarded as a valuable natural resource, and there is much concern among academic and industrial researchers, government regulatory officials and environmental organisations on how best to manage their use. An international Bt Management Working Group has been established and has been co-ordinating relevant research since 1988 (see Koziel et al., 1993). There is, of course, much debate on the consequences of introducing toxin in commercial crops. It seems likely that highly expressing plants could in many circumstances help to accelerate resistance development, and many experts (McGaughey & Whalon, 1992; Tabashnik, 1994) argue that the plants should only be used within a wider strategy of Integrated Pest Management. Such a strategy would involve the use of more than a single toxin or control agent, reducing the selection pressure exerted by each one, and maintaining also an adequate supply of susceptible breeding individuals, possibly by providing refuges for them. Within such a strategy, transferred Bt genes could be deployed in a number of ways. They may be present singly or multiply within a crop plant, and may be expressed constitutively or in a tissue-, temporal- or wound-specific manner, they may be expressed to different degrees, and the transformed plants can be used alone, in mixtures or in rotations. There are many variables, and the most useful combination will depend on the crop, the environment, and the homogeneity and behaviour of the pest populations. In a US context, McGaughey & Whalon (1992) urge the need for organised strategies and their continual evaluation to conserve the usefulness of Bt toxins for as long as possible.

Interest in transgenic plants containing Bt toxins, especially cotton plants, is world wide. Much of the investment in the relevant research has been in the USA. There is related research in Europe,often with strong links with US industrial and academic concerns: this is true, for example, of a project at the Horticultural Research Institute concerned with introducing Bt toxin into apple trees to confer resistance to the codling moth. The HRI also holds a "library" of over 6000 strains of Bt whose toxicity is being assessed and, where possible, being improved by techniques such as plasmid manipulation (e.g. Jarrett & Stephenson, 1990). The research program does not involve transforming plants. A larger project in Cambridge University is concerned with characterising new natural and engineered toxins and establishing their mechanisms of toxicity. The search for new Bt toxins with different specificities is continuing with enthusiasm world wide (Feitelson et al., 1992), and many companies have large collections of Bt strains from world wide sources. Enthusiasts (Feitelson et al., 1992) argue that the characterisation of new toxins, and also their laboratory modification, is an important approach to combat developing resistance. Their use, however, even when it involves genetic manipulation, may not necessarily involve manipulating plants. The toxins could be transferred, as are some established Bt toxins, into bacteria which when killed without being disrupted, can be used as crop sprays; alternatively they can be transferred into harmless endophytic bacteria that can be used to inoculate crop plants.

Patent claims (see Feitelson et al., 1992:Koziel et al., 1993) suggest that Bt toxins active against nematodes, flatworms and protozoa have been identified. Feitelson et al., (1992) optimistically suggest that "it may be possible to find Bt strains specific to almost any pest target from single cell eukaryotes to the most advance arthropods". It is difficult to judge these claims from current publications.

7.3   RESISTANCE BASED ON PROTEINASE-INHIBITORS.

Many plants contain proteins whose sole known activity is the inhibition of proteolytic enzymes such as the animal digestive enzymes, trypsin and chymotrypsin. They occur in leaves, seeds and fruits and may be produced constitutively or induced as the result of physical wounding, including herbivory. Their probable function is to deter herbivores by interfering with their digestion, and also to inhibit proteinases secreted by invasive fungi and bacteria. Thus Gatehouse *et al.*, (1979) argued that it was the presence of a trypsin-inhibitor in cultivars of cowpea (CpTI) that gave protection against such pests as bruchid beetles. Although the content of CpTI may not explain all the aspects of resistance in the field (Xavier-Filho 1991), Gatehouse and others have shown the toxic effects of these inhibitors in artificial diets against a range of insects (see Gatehouse *et al.*, 1991a, Burgess *et al.*, 1994).

The effect of proteinase inhibitors (PI's) on insects are most certainly more complex than simply inhibiting digestion, leading to deficiencies in essential amino acids, and growth deficiencies. They have been shown to lead to a hyperproduction of digestive enzymes (Broadway and Duffey, 1986), and it has been hypothesised that they influence water balance (Boulter, 1993) and may also affect insect moulting (Hilder *et al.*, 1993). The antinutritional effects of PI's such as hypertrophy and hyperplasia on mammals and other animals is well known (Liener, 1994). These toxicities, are however, readily destroyed by heat, and in the normal course of domestic cooking the PIs are denatured and converted into nutritionally useful sources of amino acids. The toxicity of PIs towards bacteria (see section 9), fungi (see section 8) and molluscs (see section 6) have also been reported.

PIs have been isolated from plants which are active against the four major groups of proteinases, that is the serine, cysteine, metallo and aspartic acid proteinases (see Laskowski and Kato, 1980, Ryan, 1990), but especially against the first three groups. They are generally single chain proteins encoded by single genes, and act as competitive inhibitors which bind to the active sites of proteinases. As a result they are specific for a particular class of proteinases. Within each group several distinct subgroups are recognised based on chemical and physiological characteristics. The best known of the PIs active against serine proteinases are the Kunitz and Bowman-Birk families. The toxicity of a particular PI against insects will clearly depend on the proteinases present in the digestive tract of the insect and the conditions in which they operate. PIs may therefore be expected to be rather specific against a small range of pests, and 'broad spectrum' resistance may require the presence of PIs of different types (see McManus, *et al.*, 1994).

Hilder *et al.*, (1987) demonstrated that a transferred PI gene could confer insect resistance to a transformed plant. The gene for CpTI was introduced into tobacco, and CpTI accumulated in amounts of up to 1 % of soluble protein, making the plant resistant to a commercially important lepidopteran, tobacco budworm (*Heliothis virescens*).

Resistance, measured as leaf damage and insect survival, increased with the amount of CpTI produced, and CpTI accumulation had no obvious adverse effect on the plant. CpTI has also been introduced into other crops with some success. James (1992) transformed both apple and strawberry plants, although only apple showed increased resistance and was less susceptible to predation by a range of lepidopteran and coleopteran pests. Transformations of strawberry with CpTI has also been reported by Graham (1992). Workers at Axis Genetics (Cambridge), a company that holds a patent on the CpTI gene, have transformed a number of other crops including oil seed rape, tomato, potato and lettuce. Trials suggested the plants had acquired little insect-resistance, and this was attributed to the low level of expression of the CpTI gene. Field trials of transgenic tobacco showed some resistance to the larvae of *Helicovarpa zea*, although this was less than that of tobacco transformed with a Bt toxin (Hoffmann *et al.*, 1992).

Other inhibitors of serine proteinases have been introduced into transgenic plants. Johnson *et al.*, (1989) transferred genes for tomato proteinase inhibitors I and II, as well as potato inhibitor II, into separate tobacco plants. The transgenic plant expressing proteinase inhibitor II, which inhibits the activity of trypsin and chymotrypsin, had significantly raised toxicity against *Manduca sexta* (tobacco hornworm) which increased with expression levels in plant tissue. However, the introduced tomato proteinase inhibitor I which strongly inhibits chymotrypsin activity, had little effect on the growth of larvae. The same gene for tomato proteinase inhibitor I has also been expressed in tobacco, alfalfa, and nightshade (*Solanum nigrum*), (Narváez-Vásquez *et al.*, 1992), but toxic activity was not reported. Further, the gene for potato proteinase II has also been transferred to tobacco where its expression showed an adverse effect on the growth of the Lepidopteran pest *Chysodeixis eriosoma* (green looper) (McManus *et al.*, 1994).

There is currently interest in transforming plants to produce PIs of the cysteine proteinases. These proteinases act in mildly acidic conditions, so that their inhibitors would be expected to have a different insect toxicity to those of serine proteinases. Moreover, since higher animals are thought not to utilize cysteine proteinases for digestive purposes, their inhibitors will have little or no mammalian toxicity. Feeding trials using bacteria expressing a gene for the PI oryzacystatin-I (OC-1) from rice plants, demonstrated its toxicity to a number of rice pests (Chen *et al.*, 1992). Expression of OC-1 has also been achieved in tobacco plants, (Masoud *et al.*, 1993).

The potential usefulness of transgenic plants expressing PIs has been assessed and reviewed by Boulter (1993), Gatehouse *et al.*, (1994), and Ryan (1990) among others. However, although gene expression, stable inheritance and insect resistance has been amply demonstrated, there are few reports of field trials on these plants. Their usefulness is restricted by the need for high levels of gene expression and the narrow specificity of the induced resistance. Insect resistance could be amplified and extended by incorporating genes for more than one PI into a plant, although there is no report of this having been done. PI genes have, however, been introduced together with other genes including those for lectins and Bt toxin, with encouraging results. Boulter *et al.*, (1990) crossed a CpTI-producing tobacco plant with one producing pea lectin: the progeny showed additive resistance towards attack by *H. virescens*. Similar, tobacco plants containing a trypsin inhibitor from *Cucurbita maxima* and a Bt toxin were made by MacIntosh *et al.*, (1990) using a double transformation technique. When the introduced genes both confer toxicity to a specific pest, the plants resistance will be expected to be more durable under field conditions.

7.4   RESISTANCE BASED ON LECTINS

Lectins are carbohydrate-binding proteins, other than enzymes or immunoglobulins, that bind reversibly to sugar moieties on glycoproteins, glycolipids and polysaccharides. They have been isolated from plants, animals and micro-organisms, and are classified mainly by their ability to bind specifically to particular sugar residues such as mannose or N-acetyl glucosamine (see Chrispeels & Raikhel, 1991). Within each group, the corresponding genes contain a high nucleotide homology suggesting related ancestries. Small differences in genes and in protein structures, however, ensure a variation in binding specificity and affinity.

Lectins are thought to be ubiquitous within the plant kingdom, being particularly concentrated in storage organs. Levels are especially high in leguminous and graminaceous seeds, from which sources they have been most intensively studied. They are probably part of the plants defence against herbivores and pathogens, as they are toxic to viruses, bacteria, nematodes, insects and mammals. Their toxicity to animals is presumed to result from their binding to glycoproteins on the gut wall, so causing problems in digestion, nutrient uptake and growth (Sharon & Lis, 1989). The specificity of the toxicity, which varies enormously from one lectin to another, presumably depends on the binding affinity towards particular glycoproteins, and also on the conditions inside the animal digestive system. Lectins which bind to chitin residues may also have specific effects on the chitinous membranes and cell walls of insects and fungi (Chrispeels & Raikhel, 1991). Cole (1994) at HRI (Wellsbourne), has recently described a lectin from wild, aphid-resistant *Brassica* species which probably binds to the foregut of the cabbage aphid (*Brevicoryne brassicae*) as well as to its food and salivary canals. Toxicity of lectins is, of course, destroyed by heating such as during conventional food preparation.

Feeding tests have shown individual lectins to be insecticidal to many groups of insect pests: from many orders, Coleoptera (Gatehouse *et al.*, 1991b; Cavalieri *et al.*, 1991), Homoptera (Powell *et al.*, 1993) and Lepidoptera (Cavalieri *et al.*, 1991). Some, such as the lectins from the winged bean, soyabean and wheat germ, also have mammalian toxicity. Of more interest are the lectins from peas, snowdrop and garlic which have little mammalian toxicity, probably because they are readily digested in mammalian stomachs. The snowdrop (*Galanthus sp.*) lectin (GNA), has the additional interest that it is toxic to phloem-feeding (homopteran) pests, including aphids. These pests not only damage plants but notoriously spread viruses among potato and cereal crops. Thus the attempts to introduce lectins into crop plants by genetic manipulation have mainly used these lectins of low mammalian toxicity (Boulter *et al.*, 1990; Cavalieri *et al.*, 1991; Shi *et al.*, 1994).

A major, pioneering contribution to this work was done by a group at Durham University, aided by links with industry (Axis Genetics, Cambridge and Shell Research,

Sittingbourne). Edwards (see Boulter, 1993), first transformed tobacco plants with a chimaeric pea lectin gene, CaMV 35S/P-Lec. These plants were subsequently crossed with transformed tobacco containing the cowpea trypsin inhibitor, CpTI (Boulter *et al.*, 1990). All plants expressing the lectin gene showed enhanced protection against the tobacco budworm whether this was measured as leaf damage or insect survival: in the doubly transformed plants this protection was additive to that conferred by CpTI. These plants, under the influence of the CaMV promoter, produced lectins systemically in all tissues. When however, this promoter was replaced by one from the gene for the small subunit of the photosynthetic enzyme Rubisco, lectin was produced only in photosynthetic tissues (Edwards *et al.*, 1994). This work is being developed commercially (Nickerson Biochem, Cambridge) and is expected to result in marketable plants in the foreseeable future.

Recently, the expression of lectins has been directed even more specifically at phloem-feeding insects by the use of phloem-specific promoters. Shi *et al.*, (1994) produced tobacco plants which expressed the snowdrop lectin under the control of a promoter derived from the gene for the phloem-bound enzyme, sucrose synthase-1 from rice. In these, GNA was expressed solely in the phloem of the transformed plants, and could indeed be recovered in the honeydew of peach potato aphids (*Myzus persicae*) which fed on them. Preliminary result suggest that the GNA-expressing plants limit the fecundity of predatory aphids and so control population increase. Currently rice, a crop prone to attack by phloem-feeding insects, is being transformed with this promising gene construct.

Proteins referred to as lectin-like proteins have also been identified from insect-resistant, wild strains of *Phaseolus vulgaris* where they probably confer resistance against storage pests. One of these, arcetin, is a major storage protein which replaces the more common phaseolin. Another is the powerful inhibitor of α-amylase discussed below (Moreno & Chrispeels, 1989). Both have been introduced into transgenic plants. Any insect resistance that they confer will be of interest, although Gatehouse *et al.*, (1992) express reservations about it being powerful or practically useful.

7.5   RESISTANCE BASED ON INHIBITORS OF α-AMYLASES

The inhibitors of α-amylases (α-AI), like the inhibitors of proteinases, affect major digestive enzymes of animals. Like the PIs they are widespread in plants, have different spectra of inhibitory activity and can be insecticidal in feeding tests. Like the PIs also, they have been introduced transgenically into plants and shown to confer a measure of resistance against specific insect pests.

α-AIs occur abundantly in the grains of wheat and barley (Gutierrez *et al.*, 1993), but are also present in other cereal grains and legumes including rice (Feng *et al.*, 1991), maize (Chen *et al.*, 1992) and beans (Huesing *et al.*, 1991). They are not a homogeneous group of proteins. Some have structural and biochemical similarities to cereal trypsin inhibitors (Garcia-Olmedo *et al.*, 1987) and inhibit both types of hydrolases. Others have sequence homology with lectins (Moreno *et al.*, 1990), while a protein from the graminaceous Job's tears (*Coix lachryma-jobi*) inhibits α-amylases and is also an endochitinase (Ary *et al.*, 1989). Their diversity of structure explains the differences in inhibitory specificity towards α-amylases, such that some inhibit mammalian and insect enzymes while others are specific for insect enzymes (see Gatehouse *et al.*, 1992). Most have been characterised by their activity against specific enzymes *in vitro*. The experience of Gatehouse *et al.*(1992) warns that activity against an insect enzyme does not necessarily imply insecticidal activity *in vivo*. Nevertheless, there is evidence that α-AIs, of *Phaseolus vulgaris* in particular, are correlated with resistance to such storage pests as *Callosobruchus chinensis* and *Zabrotes subfasciatus*, and when purified they are toxic to these organisms in feeding trials (see Gatehouse *et al.*, 1992).

Several groups of workers have produced transgenic plants containing raised levels of α-AI. Most notably, Altabella & Chrispeels (1990) constructed a chimeric gene from a seed specific promoter linked to a gene for an α-AI from *Phaseolus vulgaris*. This inhibitor is the one referred to above which is also recognised as a lectin, phytohaemagglutinin-2. Seeds of tobacco plants transformed with this gene contain the expected protein, and extracts strongly inhibit α-amylase from the midgut of *Tenebrio molitor* beetles. Recently Shade *et al.*, (1994) and Schroeder *et al.*, (1995) have transformed peas with this gene. The seeds were shown to contain α-AI at levels up to 1.3% of their soluble protein, and to be resistant to three bruchid beetle pests, the cowpea (*Callosobruchus maculatus*), the azuki bean (*Callosobruchus chinensis*) and the pea (*Bruchus pisorum*) weevils.

Although these results are promising, plants transformed with α-AIs have not yet been developed or tested to the same degree as those transformed with PIs. However, insects may well develop resistance mechanisms that will operate against single, unsupported α-AI genes. Storage pests feeding on seeds containing high levels of α-AIs produce compensatory high levels of α-amylase (Silano, 1975). Moreover, the wheat pest *Tribolium confusum*, unlike *Callosobruchus maculatus* which is not a pest of wheat, appears to be able to detoxify the α-AI of wheat to an extent that gives it a great degree of tolerance (Gatehouse *et al.*, 1992).

7.6   GENETIC MANIPULATION OF SECONDARY METABOLISM TO PRODUCE INSECT TOXINS

Many plant secondary metabolites have a broad spectrum of toxicity to organisms including insects (Harborne & Baxter, 1993). It is possible that in the future these will form the pool from which new toxicants, capable of being introduced into crop plants by genetic manipulation, will be selected. However, it will be a difficult task to identify and manipulate all the genes necessary for a "foreign" biosynthetic pathway. Defensive compounds could be produced from established biosynthetic pathways in one of two ways (Pickett, 1985). The first is to increase the synthesis of a suitable toxic compound that is already made in very small amounts, by enhancing the activity of the rate-limiting enzymes in its synthesis. This could be done specifically in the vulnerable tissues of the plant. The second way is to modify an existing pathway by inserting a gene to produce a new toxic compound. This aim has been achieved against fungal pathogens by Hain *et al* (1990) who transferred the gene for stilbene synthase from grape into tobacco. The tobacco plant was enabled to produce the phytoalexin resveratrol from common precurors, and so acquired resistance to the pathogen *Botrytis cinerea*.

56

For such an approach to work against insects, it requires detailed information on the toxicity of specific components and of their biosynthetic pathways. Many toxicants, such as the cyanogenic glucosides, have undesirably unspecific toxicity. Similarly, selected toxicants, such as the pyrethrins from the pyrethrum daisy (*Tanacetum cinerariaefolium*) and azadiractin from the Indian neem tree (*Azadiracthta indica*), have complex structures which are the result of complex biosynthetic pathways. Inspite of careful consideration and of a commercial claim concerning pyrethrins, they do not appear to be suitable candidates for genetic manipulation. There are comprehensive lists of plant secondary products and their toxicity (e.g. Harbourne & Baxter, 1993): selecting suitable substances for manipulation, depends on scientific imagination and ingenuity.

The usefulness and safety of a toxicant would, of course, be enhanced if its synthesis were restricted to specific vulnerable tissues, and, possibly, if its synthesis was under the control of wound-induced signals within the plant such as salicylic acid or jasmonic acid (see Bennett & Wallsgrove, 1994).

7.7 GENETIC MANIPULATION OF SECONDARY METABOLISM TO PRODUCE SEMIOCHEMICALS

The behaviour of insects in colonizing, feeding and reproducing on particular plants is determined by many non-toxic products of secondary metabolism. The presence of feeding attractants and repellants among the non-nutrient, lipophilic components of leaves has been known for a long time. More recent interest has centred on the volatile components that attract insect pests, and on those components that resemble, or are intermediates of, the behaviour-controlling insect pheromones. These signalling chemicals, or semiochemicals, could be manipulated to make a crop less attractive to insect pests. But, more radically, the manipulated, "unattractive" crop plants, could be used in conjunction with manipulated "lure" crops that had been made more attractive, and which, along with the attracted insects, could be sprayed with insecticide or insecticidal pathogens. This strategy of insect control, the "push-pull" or "stimulo-deterrent diversionary strategy" (SDDS) has been described in detail and advocated in a number of reviews (Miller & Cowes, 1990; Pickett *et al.*, 1991). Enough work has been done to show the feasibility of different aspects of this strategy. It has attracted interest internationally, and in the UK for instance, the necessary background work is being conducted, with overseas collaboration, at Research Institutes (IACR-Rothamsted; IHR-Wellesbourne) and Universities (Imperial College, Silwood Park).

Exploitation of this approach depends upon knowing the volatile semiochemicals to which insects respond. This can be established using sophisticated analytical systems in which the volatile fractions of plants are separated into their components by gas chromatography (GC), their structures determined by mass spectrography (MS), and their activity on a particular insect sensory organ be detected electrophysiologically (Wadhams, 1990; Nottingham *et al.*, 1991; Campbell *et al.*, 1993). The sensitivity of this technique is such as to establish from among the 300 volatile compounds emanating from oilseed rape, that it is two alkenyl isothiocyanates which specifically attract the cabbage seedpod weevil (*Ceutorhynchus assimilis*) (Blight *et al.*, 1992; Smart *et al.*, 1993). This suggests that rape, manipulated genetically to produce less of these isothiocyanates while maintaining its level of related glucosinolates, will be less attractive to the weevil.

Similarly the GS-MS-electrophysiological techniques have identified methyl salicylate in the aromatic components of the bird cherry tree, as a deterrent for the bird cherry-oat aphid (*Rhopalosiphum padi*) of the cereal pest stage. In field experiments,spraying this compound onto cereals decreased the summer migration and colonisation of *R. padi*, and of other aphids also, by a half (Pettersson *et al.*, 1994): this was judged sufficient to limit damage due to the cereal crop by the virus-carrying aphid.

Methyl salicylate is a volatile derivative of salicylic acid, which is known to be an endogenous chemical messenger and which helps induce some plant defence systems including the pathogenesis-related proteins (see Pierpoint, 1994; Bennett & Wallsgrove, 1994). Methyl jasmonate is another such natural volatile compound which also induces defence compounds (see Bennett & Wallsgrove 1994) and may indeed be a signal compound by which damaged plants induce resistance mechanisms in neighbouring plants (Farmer & Ryan 1990). Thus the chemical signals by which a damaged leaf "alerts" undamaged, neighbouring leaves, can be related to those that affect insect behaviour. Elucidation of these interaction mechanisms will almost certainly indicate novel mechanisms by which the secondary metabolism of a plant may both induce protective mechanisms and deter potential insect predators.

A striking example of the interdependence and co-evolution of plants and insects, is that some cyclised terpenes, characteristic of plants of the *Nepeta* genus (Labiatae), are also aphid pheromones. These compounds, especially nepetalactone and nepetalactol, acting singly or in combination, govern the attraction and mating behaviour of aphids (see Pickett *et al.*, 1992). Characterization of the four or five enzymes involved in synthesizing nepetalactone from the ubiquitous metabolite, geranyl pyrophosphate, is underway, aided by the fact that similar enzymes, involved in the synthesis of indole alkaloids, have been well studied (de Luca, 1993). Thus, these *Nepeta* plants, including the catmints (*N.racemosa* and *N.cataria*) should form a valuable source of genes which can be used not only for the "clean" synthesis of these pheromones (Hallahan *et al.*, 1992), but in producing pheromone-producing plants that could act as aphid-attracting lures. Such lure plants should be doubly effective; not only should they attract aphids, but they should also attract wasps of the genus *Praon* that are parasitoids of the aphids (Hardie *et al.*, 1994). Other examples of using semiochemicals to attract predators or parasitoids are given by Bjostad *et al.*, (1992).

7.8   CONCLUSIONS

Crop plants can be genetically manipulated to produce insecticidal toxins that confer a satisfactory, hereditary stable resistance against target insects.

Most experience concerns plants transformed to produce toxins from the bacterium *Bacillus thuringiensis*, which have insignificant mammalian toxicity. Such transformation is pre-eminently suited to crops such as cotton, which are highly susceptible to rapidly-multiplying insects, and which otherwise require frequent spraying with Bt toxins or other insecticides. Engineering plants to produce the toxins has advantages over spraying the toxins; nevertheless, it is likely that resistance-breaking strains of insects will evolve unless the toxins are supported with other resistance mechanisms, or used within the context of an integrated system of pest control. Work on Bt toxins attracts commercial support: public support is perhaps best directed at fundamental aspects. Other toxins which have been introduced to produce specific resistances include inhibitors of insect digestive enzymes, such as proteinase inhibitors and α-amylase inhibitors, and lectins. Possibly the most interesting of these is the lectin from snowdrop, which can be expressed exclusively in phloem-tissue, and which is active against phloem-feeding aphids which are major vectors of viruses. To be most effective, the genes for these proteins may need to be modified, as have those for the Bt toxins. The expertise to select and to modify these genes is well established.

It should be possible to manipulate plants to modify the production of secondary metabolites, such as the glucosinolates of brassicas, that repel specific insects. Proper testing and exploitation of this possibility requires a better understanding of the role of these substances in plant protection and of their biosynthetic pathways. A longer term, publically acceptable strategy for pest control could be based on the manipulation of the volatile signalling chemicals, the semiochemicals, that influence insect behaviour. Basic research aimed at identifying these chemicals, characterising their activity and manipulating their production is established internationally.

7.9   REFERENCES

Adang, M.J., Brody, M.S. & Caedineau, G. *et al.*, (1993). The reconstruction and expression of a *Bacillus thuringiensis cry111A* gene in protoplasts and potato plants. *Plant Molecular Biology*, **21**, 1131-1145.

Altabella, T. & Chrispeels, M.J. (1990). Tobacco plants transformed with the bean αai gene express an inhibitor of insect α-amylase in their seeds. *Plant Physiology*, **93**, 805-810.

Ary, M.B., Richardson, M. & Shewry, P.R. (1989). Purification and characterisation of an insect α-amylase inhibitor/endochitinase from seeds of Job's Tears (*Coix lachryma-jobi*). *Biochimica et biophysica acta*, **993**, 260-266.

Bennett, R.N., & Wallsgrove, R.M. (1994). Secondary metabolites in plant defence mechanisms. *New Phytologist*, **127**, 617-633.

Bjostad, L.B., Hibbard, B.E & Cranshaw, B.S. (1992). Applications of semiochemicals in integrated pest management programs. In *Pest Control with Enhanced Environmental Safety*. ACS Symposium Series, **524**, 199-218.

Blight, M.M., Hick, A.J., Pickett, J.A. *et al.*, (1992). Volatile plant metabolites involved in host plant recognition by the cabbage seed weevil, *Ceutorhynchus assimilis*. In *Proceeding 8th International Symposium of Insect-Plant Relationships*. (S.B.J.Menken, J.H.Visser and P. Harrewijn eds), pp. 105-106. Kluwer Academic Publications; Dordrecht.

Boulter, D., Edwards, G.A., Gatehouse A.M.R. *et al.*, (1990). Additive protective effects of different plant-derived insect resistance genes in transgenic tobacco plants. *Crop Protection*, **9**, 351-354

Boulter, D. (1993) Insect pest control by copying nature using genetically engineered crops. *Phytochemistry*, **34**, 1453-1466.

Broadway, R.M. & Duffey, S.S. (1986). Plant proteinase inhibitors: Mechanism of action and effect on the growth and digestive physiology of larval *Heliothis zea* and *Spodoptera exiqua*. *Journal of Insect Physiology*, **32**,827-833

Burgess, E.P.J., Main, C.A., Stevens, P.S. *et al.*, (1994). Effects of the proteinase inhibitor concentration and combinations on the survival, growth and gut enzyme activities of the black field cricket, *Teleogryllus commodus*. *Journal of Insect Physiology*, **40**, 803-811

Campbell, C.A.M., Petterson, J., Pickett, J.A.(1993). Spring migration of damson-hop aphid, *Phorodon humuli*, and summer host plant derived semiochemicals released on feeding. *Journal of Chemical Ecology*, **19**, 1569-1576.

Cavalieri, A., Czapla, T. & Howard, J. (1991). Larvicidal lectins and plant insect resistance based thereon. European patent application 90312171.3. Publication number 0 427 529 A1

Chen, M.S., Johnson B., Wen, L., *et al.*, (1992). Rice cystatin: Bacteria expression, purification, cysteine proteinase inhibitory activity and insect growth suppressing activity of a truncated form of the protein. *Protein Express. Purif.* **3**, 41-49

Chen, M.S., Feng, G.,Zen, K.C. *et al.*, (1992). α-Amylases from three species of stored grain coleroptera and their inhibition by wheat and corn proteinaceous inhibitors. *Insect Biochemistry and Molecular Biology*. **22**, 261-268

Chrispeels, M.J. & Raikhel, N.V. (1991). Lectins, lectin genes, and their role in plant defense. *The Plant Cell*, **3**, 1-9

Cole, R.A. (1994). Isolation of a chitin-binding lectin, with insecticidal activity in chemically-defined synthetic diets, from two wild brassica species with resistance to cabbage aphid *Brevicoryne brassicae*. *Entomologia experimentalis applicata*, **72**, 181-187

de Luca, V. (1993). Enzymology of indole alkaloid biosynthesis. *Methods in Enzymology*, **9**, 345-368.

Edwards, G.A., Hepher, A., Clerk, S.P. *et al.*, (1994). Pea lectin is correctly processed, stable and active in leaves of transgenic potato plants. *Plant Molecular Biology*, **17**, 89-100

Farmer, E.E. & Ryan, C.A. (1990). Interplant communication: airborne methyl jasmonate induces synthesis of proteinase inhibitors in plant leaves. *Proceedings of the National Academy of Science*, **87**, 7713-7716.

Feitelson, J.S., Payne, J. & Kim, L. (1992). *Bacillus thuringiensis*: Insects and beyond. *Bio/Technology*, **10**, 271-275.

Feng, G.H., Chen, M., Kramer, K.J. *et al.*, (1991). α-Amylase inhibitors from rice. *Cereal Chemistry*, **68**, 516-521.

Ferre, J. Real, M.D. Rie J.V. *et al.*, (1991). Resistance to the *Bacillus thuringiensis* bioinsecticide in a field population of *Plutella xylostella* is due to a change in a midgut membrane receptor. *Proceedings of the National Academy of Science of the USA*, **88**, 5119-5123.

Fischoff, D.A., Bowdish, K.S., Perlak, F.J. *et al.*, (1987). Insect tolerant transgenic tomato plants. *Bio/Technology*, **5**, 805-812.

Fraley, R. (1992) Sustaining the food supply. *Bio/Technology*, **10**, 40-43.

Garcia-Olmedo, F., Salcedo, G., Sanchez-Monge, R. *et al.*, (1987). Plant proteinaceous inhibitors of proteinases and α-amylases. In *Oxford Surveys of Plant Molecular, Cell Biology*, **4**, 275-334.

Gasser, C.S. & Fraley, R. (1989) Genetically engineering plants for crop improvements. *Science*, **244**, 1293-1299.

Gatehouse, A.M.R., Boulter, D. & Hilder, V.A. (1991). Novel insect resistance using proteinase inhibitor genes. In *Molecular Approaches to Crop Improvement* (E.S. Dennis and D.J. Llewellyn eds), pp 63-77. Springer-Verlag: New York.

Gatehouse, A.M.R., Howe, D.S., Flemming, J.E. *et al.*, (1991). Biochemical basis of insect resistance in winged bean seeds (*Psophocarpus tetragonolobus*) seeds. *Journal of the Science of Food and Agriculture*, **55**, 63-74

Gatehouse, A.M.R., Gatehouse, J.A., Dobie, P. *at al* (1979). Biochemical basis of insect resistance in *Vigna unguiculata*. *Journal of the Science of Food and Agriculture*, **30**, 948-958

Gatehouse, A.M.R. & Hilder, V.A. (1994). Genetic manipulation of crops for insect resistance. In *Molecular Biology in Crop Protection* (G. Marshall and D. Waters eds), pp 177-201 Chapman and Hall: London.

Gatehouse, A.M.R. Boulter, D. & Hilder, V.A.(1992). Potential of plant derived genes in the genetic manipulation of crops for insect resistance. In *Plant Genetic Manipulation for Crop Protection* (A.M.R. Gatehouse, V.A. Hilder and D. Boulter eds), pp 155-181. C.A.B. International: Wallingford UK

Gibbons, (1991). Moths take to the field against biopesticide. *Science*, **254**, 646.

Goldberg, R.J. & Tjaden, G. (1990). Are B.T.K. plants really safe to eat? *Bio/Technology*, **8**, 1011-1014.

Graham, J. (1992). Applications of biotechnology to soft fruit breeding. *SCRI Annual report 1992*, 23-25

Gutierrez, C., Garcia-Casado, G.C., Sanchez-Monge, *et al.*, (1993). Three inhibitor types from wheat endosperm are differentially active against α-amylases of Lepidoptera pests. *Entomologia experimentalis applicata*, **66**, 47-52.

Hain, R., Reif, H.J, Krause, E. *et al.*, (1993). Disease resistance results from foreign phytoalexin expression in a novel plant. *Nature*, **361**, 153-156.

Hallahan, D.L. Pickett, J.A. Wadhams, L.J. *et al.*, (1992). Potential of secondary metabolites in genetic engineering of crops for resistance. In *Plant Genetic Manipulation for Crop Protection* (A.M.R. Gatehouse, V.A. Hilder and D. Boulter eds), C.A.B. International; Wallingford.

Harborne, J.B. & Baxter, H. (1993). Eds. Phytochemical Dictionary: a Handbook of Bioactive Compounds from Plants. Taylor and Francis; London'

Hardie, J., Hick, A.J., Holler, C. *et al.*, (1994). The response of *Praon* spp parasitoids to aphid sex pheromone compounds in the field. *Entomologia experimentalis applicata*, **71**, 95-99.

Hilder, V.A., Gatehouse, A.M.R., Sheerman, S.E. *et al.*, (1987). A novel mechanism of insect resistance engineered into tobacco. *Nature*, **300**, 160-163

Hilder, V.A., Gatehouse, A.M.R. & Boulter, D. (1993). Transgenic plants conferring insect tolerance, protease inhibitor approach. In *Transgenic Plants Vol. 1* (S.D. Kung and R. Wu eds), pp. 317-338 Academic Press: San Diego.

Hoffmann, M.P., Zalom, F.G., Wilson, L.T. *et al.*, (1992). Field evaluation of transgenic tobacco containing genes encoding *Bacillus thuringiensis* δ-endotoxin or cowpea trypsin inhibitor: Efficacy against *Helicoverpa zea* (Lepidoptera: Noctuidae). *Journal of Economic Entomology*, **85**, 2516-2522

Huesing, J.E., Shade, R.E. Chrispeels, M.J. *et al.*, (1991). α-Amylase inhibitor, not phytohemagglutinin, explains resistance of common bean seeds to cowpea weevil. *Plant Physiology*, **96**, 993-996.

James, D.J., Passey, A.J., Easterbrook, M.A. *et al.*, (1992). Transgenes for pest and disease resistance: Progress in the introduction of transgenes for pest resistance in apples and strawberries. *Phytoparasitica* **20**, Suppl. 83S-87S

Jarrett, P. & Stephenson, M. (1990). Plasmid transfer between strains of *Bacillus thuringiensis* infecting *Galleria mellonella* and *Spodoptera littoralis*. *Applied Envirnmental Microbiology*, **56**, 1608-1614.

Johnson, R., Narvaez, J., Gynheung, A. *et al.*, (1989). Expression of proteinase inhibitors I and II in transgenic tobacco plants: Effect on natural defence against *Manduca sexta* larvae. *Proceedings of the National Academy of Science of the USA*, **86**, 9871-9875

Koziel, M.G., Carozzi, N.B., Currier, T.C. *et al.*, (1993) The insecticidal crystal proteins of *Bacillus thuringiensis*: Past, present and future uses. *Biotechnology Genetic Engineering Reviews*, **11**, 171-228.

Laskowski, M. & Kato, I. (1980). Protein inhibitors of proteinases. *Annual Review of Biochemistry*, **49**,593-626

Liener, I.E. (1994). Implications of antinutritional components in soybean foods. *Critical Reviews of Food Science Nutrition*, **34**, 31-67

MacIntosh, S.C., Kishore, G.M., Perlak, *et al.*, (1990). Potentiation of *Bacillus thuringiensis* insecticidal activity by serine proteinase inhibitors. *Journal of Agricultural and Food Chemistry*, **38**, 1145-1152

Masoud, S.A., Johnson, L.B., White, F.F. *et al.*, (1993). Expression of a cysteine proteinase inhibitor (oryzacystatin-I) in transgenic tobacco plants. *Plant Molecular Biology*, **21**,655-663

McGaughey, W.H. & Whalon, M.E. (1992). Managing insect resistance to *Bacillus thuringiensis* toxins. *Science*, **258**, 1451-1455.

McManus, M.T., White, D.W.R. & McGregor, P.G. (1994). Accumulation of a chymotrypsin inhibitor in transgenic tobacco can affect the growth of insect pests. *Transgenic Research*, **3**, 50-58

Miller, J.R. & Cowes, R.S. (1990). Stimulo-deterrent diversion: a concept and its possible applications to onion maggot control. *Journal of Chemical Ecology*, **16**, 3197-3213.

Moreno, J., Altabella, T. & Chrispeels, M.J. (1990). Characterization of α-amylase inhibitor, a lectin-like protein in the seeds of *Phaseolis vulgaris*. *Plant Physiology*, **92**, 703-709.

Moreno, J. & Chrispeels, M.J. (1989) A lectin gene encodes the α-amylase inhibitor of the common bean. *Proceedings of the National Academy of Science of the USA*, **86**, 7885-7889.

Narváez-Vásquez, J., Orozco-Cárdenas, M.L, & Ryan, C.A. (1992). Differential expression of a chimeric CaMV-tomato proteinase Inhibitor I gene in leaves of transformed nightshade, tobacco and alfalfa plants. *Plant Molecular Biology*, **20**, 1149-1157

Nottingham, S.F., Hardie, J., Dawson, G.W. *et al.*, (1991). Behavioral and electrophysiological responses of aphids to host and nonhost plant volatiles. *Journal of Chemical Ecology*, **17**, 1231-1242.

Peferoen, M. (1992) Engineering of insect-resistant plants with *Bacillus thuringiensis* crystal protein genes. In *Plant Genetic Manipulation for Crop Protection* (A.M.R. Gatehouse, V.A. Hilder and D. Boulter eds), pp 135-153. C.A.B. International. Wallingford; London.

Pettersson, J., Pickett, J.A., Pye, B.J. *et al.*, (1994). Winterhost component reduces colonization by the bird-cherry-oat aphid, *Rhopalosiphum padi*, and other aphids in cereal fields. *Journal of Chemical Ecology*, (in press).

Pickett, J.A. (1985). Production of behaviour-controlling chemicals by crop plants. *Philosophical Transactions of the Royal Society Series B London*, **310**, 235-239.

Pickett, J.A., Wadhams, L.J. & Woodcock, C.M. (1991). New approaches to the development of semiochemicals for insect control. In *Proceedings of the Conference on Insect Chemical Ecology: Tabor*, 1990 pp 333-345. Academia, Prague and SPB Acad.Publ., The Hague.

Pickett, J.A., Wadhams, L.J., Woodcock C.M. *et al.*, (1992). Chemical ecology of aphids. *Annual Review of Entomology*, **37**, 67-90.

Pierpoint, W.S. (1994). Salicylic acid and its derivatives in plants: medicines, metabolites and messenger molecules. *Advances in Botany Research*, **20**, 164-193.

Powell, K.S., Gatehouse, A.M.R., Hilder, V.A. *et al.*, (1993). Antimetabolic effects of plant lectins and plant and fungal enzymes on the nymphal stages of two important rice pests, *Nilaparvata lugens* and *Nephotettix cinciteps*. *Entomologia experimentalis applicata*, **66**, 119-126

Ryan, C.A. (1990). Proteinase inhibitors in plants: Genes for improving defences against insects and pathogens. *Annual Review of Phytopathology*, **28**, 425-449.

Schroeder, H.E., Gollasch, S., Moore, A. *et al.*, (1995). Bean α-amylase inhibitor confers resistance to the pea weevil (*Bruchus pisorum*) in transgenic peas. *Plant Physiology*, **107**, 1233-1239.

Shade, R.E., Schroeder, H.E., Pueyo, J.J. *et al.*, (1994). Transgenic pea seeds expressing the α-amylase inhibitor of the common bean are resistant to Bruchid beetles. *Bio/Technology* **12**, 793-796.

Sharon, N. & Lis, H. (1989). Lectins. Chapman and Hall: London.

Shi, Y., Wang, M.B., Powell, K.S. *et al.*, (1994). Use of the rice sucrose synthase-1 promoter to direct phloem-specific expression of β-glucuronidase and snowdrop lectin genes in transgenic tobacco plants. *Journal of Experimental Botany*, **45**,623-631

Silano, V, Furia, M. Gianfreda, L. *et al.*, (1975). Inhibition of amylases from different origins by albumins from the wheat kernel. *Biochemica et biophysica acta*, **391**, 170-178.

Smart, L.E., Blight, M.M. & Hick, A.J. (1993). Development of a monitoring system for the cabbage seed weevil and the pollen beetle. *IOBS wprs Bulletin*, **16**, 351-354.

Tabashnik, B.E.(1994). Evolution of resistance to *Bacillus thuringiensis*. *Annual Review of Entomology*, **39**, 47-79.

Wadhams, L.J. (1990). The use of coupled gas chromatography:electrophysiological techniques in the identification of insect pheromones. In *Chromatography and isolation of insect hormones and pheromones* (A.R. McCaffery and I.D. Wilson eds), pp289-298. Plenum Press; New York.

Xavier-Filho, J. (1991). The resistance of seeds of cowpea (*Vigna unguiculata*) to the cowpea weevil (*Callosobruchus maculatus*). *Mémoires de l'Institut of Oswaldo Cruz, Rio de Janeiro*, **86**, Suppl. II, 75-77B

8.   MODIFYING RESISTANCE TO PLANT-PATHOGENIC FUNGI

W.S. Pierpoint, J.A. Hargreaves and P.R. Shewry
(IACR-Rothamsted and IACR-Long Ashton)

8.1   BACKGROUND

The techniques of genetic manipulation have not yet, to our knowledge, provided any fungal-resistant plants that are of immediate commercial use. Nevertheless, research work in this area has been vitalized by the new techniques which have not only provided, and will continue to provide, essential insights into the mechanisms of natural plant resistance, but have also provided the means by which ideas, some of which have been debated for decades, can be directly tested. A significant advance has been the isolation and partial characterization of several major resistance (R) genes of crop plants. A fuller knowledge of how they work will almost certainly be turned to practical advantage in the future, even if it is not yet clear how this will be achieved. The techniques have also led to the construction of transgenic plants, usually tobacco plants, which contain one or two introduced genes and which have a measurably increased resistance to some fungi. This work has emphasised the advantages of introducing two such genes together, as they often act synergistically and, moreover, form a defence that is likely to be less readily overcome by mutations in the pathogenic fungi. Thirdly, these techniques have produced an encouraging example of how the secondary metabolism of plants may be modified to produce a novel, functional phytoalexin. Although in practical terms, research work on fungal resistance lags behind that of work on resistance to insects or viruses, there appear to be many opportunities. The undoubted importance of damage done by fungal pathogens warrants the exploitation of these opportunities.

8.2   THE ISOLATION OF PLANT R-GENES

Techniques of chromosome mapping and of tagging genes with transposons are now well enough developed to enable the location and characterisation of the R genes which determine resistance against specific strains of pathogens. As yet the techniques are only useful in plants which have a small genome (e.g. *Arabidopsis thaliana)*, a well-mapped genome (tomato), or plants such as maize and tobacco which either have active transposons or can accept them. Moreover they require a major research effort. Nevertheless, their development is such that at a conference in Edinburgh in July (1994) the structural details of some four R genes were made public, thus increasing the total number to six. It is widely believed that this number will increase dramatically over the next few years. The techniques used and the usefulness of the information they produce has been reviewed both in detail (Bennetzen & Jones, 1992; Newbury, 1992) and more generally (Cornelissen & Melchers, 1993: Moffat, 1994; Knogge, 1994).

Of the six known R genes, the structure of five have been reported in full: two of them confer resistance to races of fungi, two to a bacterium, and one to a virus. In only one case has the encoded protein been characterised and its function established. The proteins corresponding to the other genes are unknown and their structures only inferred from amino acid sequences: nevertheless these give tantalising glimpses of interactions that must be the first in determining resistance. The one established gene product is that derived from the *Hml* gene from maize which confers resistance to the fungus *Cochliobolus (= Helminthosporium) carbonum* and was, incidentally, the first R gene to be isolated (Johal & Briggs,1992). The protein is an NADP-dependent reductase which reduces, and so detoxifies, a pentapeptide toxin which is secreted by the fungus and which is the cause of its pathogenicity. The *Hml* gene thus determines a type of resistance which is genetically and mechanistically different from that determined by the more general type of R gene, in which plant resistance rather than pathogenicity is determined by a single, pathogen-borne avirulence gene. It is this second type of resistance that follows the gene-for-gene relationship proposed by Flor (see Newbury, 1992), and which forms the highly race specific resistance that is more generally encountered.

The first representative of this type of R gene to be isolated (Martin *et al.*, 1993) was *Pto* which determines the resistance of tomato to races of *Pseudomonas syringae*. Its structure suggests that it encodes a comparatively small (321 amino acid) cytoplasmic protein that functions as a serine-threonine protein kinase. Such kinases are common in both animal and plant cells, where they may control enzymic responses by phosphorylation. They are often associated with specific recognition events such as those that occur between pollen and stigma cells. Such a protein seems admirably suited for a role in a signal-transduction pathway, but its inferred structure gives no hint of how it could interact with a component of a pathogen. However, such recognition sites can be tentatively identified in the proteins encoded by the other three R genes.

These genes are the *RPS2* from *A. thaliana* which confer resistance to *P. syringae* (Bent *et al.*, 1994: Mindrinos *et al.*, 1994), *Cf-9* from tomato that confers resistance to the fungus *Cladosporium fulvum* (Jones *et al.*, 1994), and the N gene from tobacco cvs. which triggers the hypersensitive response to TMV (Whitham *et al.*, 1994) They code for large proteins containing 909, 863 and 1144 amino acids respectively. From their amino acid sequences it is believed that they are glycoproteins, but while the *Cf-9* protein is extracellular and anchored to the plant cell membrane, and the *RPS2* protein spans the membrane and has both extracellular and intracellular domains, the N gene protein is completely intracellular. However, all three have a common motif which is composed of a number (28-4) of imperfect repeats of a leucine-rich sequence of approx 26 amino acids. This motif (LRR) occurs widely in proteins and is associated with regions that interact with other proteins and ligands. It is presumed therefore that these areas represent the receptor or recognition sites where the proteins interact directly with some specific component of the pathogen, presumably the product of the avirulence (*avr*) gene. The suggested cellular locations of the LRR regions in the three proteins are consistent with ideas on where the R gene proteins first encounter their specific pathogens; extracellularly with the fungus and bacterium, and intracellularly with the virus (Whitham *et al.*, 1994).

The structure of the *Cf-9* protein gives little indication of how, once activated at its receptor site, it could transduce a cytoplasmic signal. Speculations include the possibility that it interacts with a protein kinase such as the product of the *Pto* gene (Jones *et al.*, 1994), and indeed, it is known that the full expression of *Cf-9* requires the participation of two additional genes (Hammond-Kosack *et al.*, 1994b). In contrast, the proteins encoded by *RPS2* and N both contain intracellular regions which could generate cytoplasmic signals. Most notable are the phosphate-binding P loops, functional regions that are recognised in many proteins that bind the nucleotides ATP and GTP. The *RPS2* protein also contains a region of repeated, leucine-containing sequences, a so called "leucine zipper", that in other proteins is associated with the formation of either homo- or heterodimers. These suggested functions invite much speculation, but their existence still awaits the expression of these genes, and studies on the encoded proteins. However, the striking similarity of the *RPS2* and the N genes, make it very likely that basically similar mechanisms determine plant resistance, especially hypersensitive resistance against very dissimilar pathogens.

The isolation of these R genes represents a major step in understanding the mechanism by which plants resist pathogens. It has helped to stimulate the debate on how their manipulation could be exploited to produce new resistances in crop plants. It has yet, of course, still to be demonstrated that R genes can be transferred into other plants and be integrated functionally. Even when this can be done, it is unlikely that any race-specific resistances that they induce will be durable in the field against rapidly evolving races of fungi. However, de Wit (1992; see Cornelissen & Melchers, 1993) has already suggested how the transgenic use of the fungal avirulence gene (*avr9*) corresponding to the plant gene *Cf-9*, could be used to extend resistance in plants containing *Cf-9*. If it were introduced under the tight control of a promoter that responds rapidly and locally to a variety of pathogens, it could produce the *avr9* elicitors that interact with *Cf-9* and result in a local hypersensitive reaction. First attempts to introduce *avr9* into plants containing *Cf-9* have been made, and they provide some encouragement even though they emphasize that the constitutive expression of *avr9* can lead to a developmentally controlled, lethal necrosis in tomato seedlings (Hammond-Kosack *et al.*, 1994a). **When** this over reaction can be controlled, and **when** the *Cf-9* gene can be transgenically manipulated, there is the possibility that both *avr9* and *Cf-9* could be introduced as a resistance "gene cassette" into plants lacking *Cf9*. It would be surprising, if, in the medium-to-long term, scientific ingenuity should prove incapable of exploiting such novel opportunities for creating antifungal resistance.

8.3   RESISTANCE BASED ON ANTIFUNGAL PROTEINS

It has become clear that plants are an enormously rich source of antifungal proteins. These range from hydrolases which may normally be induced as part of a hypersensitive response to infection and other stresses, to toxic proteins, often membrane-destabilizing proteins, produced constitutively in storage tissues such as seeds and tubers. The current list of such proteins is being actively extended by wide screening programmes such as are in progress in the University of Leuven (Belgium) and in industrial research laboratories such as Zeneca in the UK. Not all these substances, of course, are suitable candidates for introducing more widely into crop plants, either because they may be toxic to mammals as are some thionins, or because they may be major allergens as are some of the 2S albumins of seeds. Nevertheless, there are many possible candidate proteins which, when suitably expressed either by themselves or together with other proteins with which they act synergistically, may introduce a useful resistance. These possibilities are being actively explored internationally.

8.3.1   Chitinases and β-1,3-Glucanases

Chitinases and β-1,3-glucanases catalyse the hydrolysis of the carbohydrate polymers chitin and β-1,3-glucan, respectively. Plant cell walls contain β-1,3-glucans but no detectable chitin, and indeed it was only recently that possible plant substrates for chitinases, i.e. cell wall glycolipids, were detected (see Collinge et al., 1993). In contrast, both polymers are major components of the cell walls of most filamentous fungi except the oomycetes. Both types of enzymes exist in a series of families or classes, each of which may contain a number of isoforms. The enzymes of each class differ from those in other classes in structure, cellular location and enzyme activity, and their synthesis may be differently regulated by developmental stimuli, exogenous hormones and pathogenic infection (see Collinge et al., 1993, 1994; Broglie & Broglie, 1993: Payne et al., 1990). Thus in the tobacco plant, where they have been well studied, there are at least three classes of each enzyme. Most relevant are the class 1 enzymes, which are basic, vacuolar proteins, and which are pathogen-induced in leaves, although they may be constitutively present in roots. These enzymes are antifungal in in vitro tests, where they inhibit the growth of Fusarium solani and Trichoderma viride among other pathogens, at concentrations as low as 10-30 μg per ml (Mauch et al., 1988: Sela-Buurlage et al, 1993).

The fungal toxicity of chitinases and glucanases is often markedly synergistic, and together they produce a characteristic swelling and lysis of growing hyphal tips. The class II enzymes have structural similarities to the class I enzymes but are acidic proteins which are secreted into intercellular spaces. However, although the induced synthesis of these intercellular enzymes is characteristic of the hypersensitive response to pathogens, they have little fungal toxicity and no synergistic interaction with class I enzymes. A third class of β-1,3-glucanases has been identified in tobacco leaves (Payne et al., 1990), and Melchers et al., (1994) have identified at least three more classes of chitinases including those from plants other than tobacco. One of these, a class IV chitinase from tobacco which resembles chitinases from soil bacteria, inhibits the growth of T. viride and Alternaria radicina and acts synergistically with class I glucanases. Collinge et al., (1993) list some 16 plants, most of them crop plants, from which genes for chitinases have been cloned.

A complex picture is emerging from the detailed and active studies being conducted on these enzymes, especially the chitinases (see Collinge et al., 1994; Broglie & Broglie, 1993). A common conclusion is that although chitinases have other roles, including some concerned with organ development, some forms of them have active antifungal roles although they are not the only factors conferring resistance even to chitinase-sensitive fungi. This view has encouraged the construction, in both academic and industrial laboratories, of transgenic plants containing these enzymes expressed constitutively, and of their examination for resistance to fungal pathogens.

These endeavours were initially rewarded when Broglie et al., (1991) transformed tobacco plants with the gene for a class I chitinase under the control of the constitutive 35S promoter from cauliflower mosaic virus. The plants developed normally and

69

expressed the enzyme so that the constitutive level of chitinase activity was 2-4 times higher in roots and 23-44 times higher in leaves. They had a quantitatively increased resistance to the soil-borne pathogen *Rhizoctonia solani* measured in terms of both seedling survival and subsequent growth rate. As expected, these transgenic plants had no increased resistance to *Pythium aphanidermatum*, a pathogen that lacks a chitin-containing cell wall (see also Broglie & Broglie, 1993). A similar resistance to *R. solani* was introduced into tobacco by Jach *et al.*, (1992) using a different class of chitinase.

The chitinase used by Jach *et al.*, (1992) was derived from the soil bacterium *Serratia marcescens* which inhibits the growth of soil fungi very effectively and has been considered as a biocontrol agent. The purpose of this transformation was not to boost the "natural" defences of the tobacco plant but to introduce what was, at the time, considered to be a novel antifungal chitinase. Experience in other laboratories was, however, less encouraging. *Nicotiana sylvestris* transformed to contain a strongly expressed gene for a class I chitinase, showed no significant resistance to the chitin-containing *Cercospora nicotianae*, the fungus responsible for frog eye disease (Neuhaus *et al*, 1991). Moreover, the same research group could find no evidence that a class I vacuolar β-1,3-glucanase contributed to resistance against the same fungus; when the activity was virtually suppressed in *N. sylvestris* by an introduced antisense gene, the plants were no more susceptible than were untransformed controls (Neuhaus *et al*,1992). These and similar results (van den Elzen *et al.*, 1993) led to the suggestions that, to be effective, these hydrolase enzymes need to be secreted into the intercellular spaces that are the niches occupied by many pathogenic fungi, or alternatively they need to be produced in synergistic combinations. The targeting sequence which directs these proteins into vacuoles rather than into intercellular spaces has now been identified and has been deleted, so that plants which secrete class I hydrolases extracellularly have been developed (Melchers *et al.*, 1993). More immediately, however, plants have been transformed to produce two hydrolases and have been found to have enhanced fungal resistance.

American workers (Zhu *et al.*, 1994) produced tobacco plants which constitutively expressed either a rice gene for a basic chitinase, or an alfalfa gene for an acidic glucanase. Both transformed plants possessed significantly more resistance to *Cercospora nicotianae* than did untransformed plants and they developed fewer, smaller lesions. Resistance was more marked, however, in heterologous plants resulting from crossing the transformants, and more emphatically so in selected progeny from these plants that were homozygous for both chitinase (2n) and glucanase (2n) genes: the degree of resistance suggested that, in doubly transformed plants, the two enzymes had synergistic antifungal activity. Resistance held up in glasshouse trials designed to imitate field conditions, and it apparently extended to the leaf spot fungus *Thanatophorus cucumeris*. Doubly transformed tomato plants have been more directly produced by Dutch workers at Mogen International (van den Elzen *et al.*, 1993), using a complex construct encoding both class I and class II chitinases and β-1,3-glucanases. The individual genes were expressed to different degrees in various transformants and the resistance of the plants to *F. oxysporum* differed. A comparison of gene expression and resistance suggests that significantly enhanced resistance requires the simultaneous expression of both class I chitinase and a class I glucanase, a result consistent with the synergistic action of these enzymes *in vitro*. Thus experience with doubly transformed plants suggests a way of enhancing resistance, and as Zhu *et al.*, (1994) emphasize, this "combinatorial" deployment of antifungal genes, especially when they direct different but complementary activities, is less likely to breakdown as a result of pathogen mutation and more likely to produce a broad, durable field resistance.

8.3.2   Other Pathogenesis-Related Proteins

Chitinases and β-1,3-glucanases are two groups of pathogenesis-related (PR) proteins which are a prominent feature of the hypersensitive response of plants to many pathogens (Bol *et al.*, 1990; van Loon *et al.*, 1994). Members of three other groups of PR-proteins have antifungal properties which may be functional and effective in infected plants.

The PR-5 group of proteins is sometimes referred to as thaumatin-like (TL) proteins because of their sequence homology with the sweet protein thaumatin from *Thaumatococcus daniellii* (Cornelissen *et al.*, 1986; Pierpoint *et al.*, 1987). Like the proteins of other PR-groups, they occur in at least two classes, basic intracellular proteins and acidic extracellular ones. The former are often called osmotins as they were first identified in osmotically-stressed plant tissue. In tobacco, as in other plants, both classes of proteins consist of a number of isoforms. Direct searches for an inhibitory activity against *Phytophthora infestans* induced in the leaves of tomato and tobacco following a hypersensitive response to a virus, showed that it resides in an intracellular PR-5 protein, osmotin II (Woloshuk *et al.*, 1991). At concentrations as low as 1 μg/ml, the purified protein caused lysis of sporangia and inhibited hyphal growth. This activity extends to a broad range of other fungi including *Candida albicans*, *Neurospora crassa* and *Trichoderma reesei*. Also, although the main extracellular PR-5 of tobacco leaves has no action on these fungi, it is active against *Cercospora beticola* (Viger *et al.*, 1992). Similar direct searching for antifungal proteins in cereal seeds including wheat, barley, oats, maize and sorghum, has revealed other inhibitory proteins with compositions similar to thaumatin (Vigers *et al.*, 1991; Hejgaard *et al.*, 1991; Huynh *et al.*, 1992). These seed proteins, zeamatin from maize, avematin from oats and trimatin from wheat etc., are collectively known as permeatins (Vigers *et al.*, 1991). Although the exact nature of their action is unknown, it results in the destabilization of hyphal membranes rendering them permeable. Thaumatin itself has a weak antifungal activity against *Candida albicans*.

Attempts are in progress to produce fungal resistant plants by introducing genes for the permeatins. A gene for the extracellular PR-5 has been introduced into tobacco (see Bol *et al.*, 1990) where it was found to be ineffective in enhancing virus resistance. In contrast, constitutive expression of osmotin in potato plants delays the development of disease symptoms after inoculation with spores of *Phytophthora infestans* (Liu *et al.*, 1994).

Recent work has also shown that PR-1 proteins are inhibitory to oomycete pathogens, either as purified proteins (to *Phytophthora infestans*) (Niderman *et al.*, 1995) or when expressed in transgenic tobacco plants (to *P.parasitica* and *Peronospora tabacina*) (Alexander *et al.*, 1993).

71

Another tobacco protein (CBP 20) of unknown function has recently been shown to be antifungal (Ponstein *et al.*, 1994). It was isolated from TMV-infected tobacco plants and identified, by amino acid sequence and immunology, as an intracellular (class I) PR-4 protein. It is active against *Trichoderma viride* and *Fusarium solani* at 6-7 µg/ ml, and it both inhibits growth and causes a lysis of germ tubes. Moreover it acts synergistically with class I chitinases and glucanases at lower concentrations, and with glucanase it inhibited the growth of *Alternaria radicina*. Again the mechanism of fungal toxicity is unknown. It is not due to chitinase activity, although the N-terminal amino acid sequence of CBP 20 is very similar to a chitin-binding domain of class I chitinases. It is striking that a chitin-binding domain that lacks chitinase activity is a characteristic feature of a number of other antifungal proteins that have been recently described. These include two basic peptides of about 30 amino acid residues from the seeds of *Amaranthus caudatus* which are active against seven tested fungi in the range of 1-10 µg/ ml (Broekaert *et al.*, 1992). Others are hevein, an antifungal protein from rubber tree latex (Parijs *et al.*, 1991) and a small chitin-binding lectin from the rhizomes of *Urtica dioica*, the stinging nettle (Broekaert *et al.*, 1989). This lectin inhibits the growth of some seven diverse chitin-containing fungi at concentrations between 20-125 µg/ml, but has no effect on the non-chitinous *Phytophthora erythroseptica*. Broekaert *et al.*, (1989) advocate its use in the genetic engineering of fungus-resistant crops.

8.3.3 Polygalacturonase-Inhibiting Protein

The endo α-1,4-polygalacturonases of fungi are important pathogenesis-determining factors: by completely degrading and solubilizing plant cell wall polygaluractans, they assist fungal colonization. In contrast, plant polygalacturonases assist plant resistance, as they produce large oligo-galacturonides which act as elicitors and induce the production of phytoalexins and other defence responses (see Toubart *et al.*, 1992 for references). Bean plants and many other dicots contain a cell wall-associated protein (PGIP) which inhibits fungal endogalacturonases but not those of plants. This protein is therefore considered to assist plant resistance by favouring the production and accumulation of the oligo-galacturonide elicitors. This potential role can clearly be tested by enhancing its production in plants or by introducing it into plants which do not produce it. A gene which encodes PGIP in beans, has been isolated and characterised (Toubart *et al.*, 1992), and is, no doubt, being used in such tests.

8.3.4 Thionins, Defensins, RIPs and Other Antifungal Proteins

There is an active international search for novel antifungal proteins from plant sources. Examples reported recently include small, basic polypeptides from the seeds of *Mirabilis jalapa* (Cammue *et al.*, 1992; De Bolle *et al.*, 1995), the 2S storage albumins from radish and from other species (Tarras *et al.*, 1992:1993a), and a class of "lipid transfer proteins" that have been characterized from a range of species (e.g. Thoma *et al.*, 1993; Molina & Garcia-Olmedo, 1993; Cammue *et al.*, 1995).

The thionins (see Florack & Stiekema,1994 for references) are cysteine-rich polypeptides of about 45 amino acid residues (~5 kDa) that were initially extracted from cereal endosperms and shown to be toxic to microorganisms. They characteristically contain

8 cysteine residues, all involved in disulphide links. The thionins from wheat and barley, called purothionins and hordothionins respectively, are best known, but over the last decade many homologous proteins have been identified both in the seeds of other species belonging to the Poaceae, and also in the leaves and stems of diverse species ranging from Abyssinian cabbage (*Crambe abyssinica*) to the mistletoe (*Viscum album*). They all contain a generally accepted consensus sequence of 18 amino acids which includes 4 cysteine residues, and they probably have similar three dimensional structures. They fall into 4 distinct types which differ in overall amino acid composition, number of S-S bonds, and overall net charge. Thionins are usually toxic to microorganisms such as yeasts, bacteria and fungi at comparatively low concentrations. They can be toxic to small animals if injected rather than ingested, and have been demonstrated to inhibit enzymes and many cellular processes that depend on intact membranes (see Florack & Stiekman, 1994). The basis for their toxicity is believed to be an interaction with membranes which makes them permeable. The interaction may initially be non-specific and depend on electrostatic charges, but it also involves a more specific interaction with inositol-containing phospholipids, possibly those involved with the transduction of signals and the control of calcium channels. This effect on membranes resembles those of some venoms and toxins produced by snakes, snails and scorpions which are also small, cysteine-rich basic proteins.

The use of thionin genes to confer resistance to plants which lack them is in its early stages of investigation. Reported results mainly concern resistance to bacteria, and these have been equivocal. Carmona and her colleagues (1993) reported that tobacco plants constitutively expressing a high level of α-hordothionin had enhanced resistance to two strains of *Pseudomonas syringae*. Infiltrated bacteria grew less well over a test period of a few days and produced smaller lesions than they did on control plants. This inhibition of bacterial growth seemed to be directly proportional to gene expression. However, similar plants produced by Dutch workers, were briefly reported to lack resistance to similar strains of *Pseudomonas* and to other bacteria (Florack & Stiekema, 1994). As some of these bacteria are known to be sensitive to cereal thionins *in vitro*, their apparent insensitivity in the plant might be due to the intercellular location of the thionins. These plants are currently being tested against fungal pathogens which are also sensitive to thionins *in vitro*, but we are unaware of the results. The synergistic enhancement of the antifungal activities of these thionins by 2S albumins or trypsin inhibitors *in vitro* (Terras *et al.*, 1993a) suggests that any fungal resistance that the transformed plants have could be enhanced by the genetic introduction of these plant proteins.

The defensins, or antifungal proteins (AFPs), are a group of small (~5kDa) proteins that were initially thought to be related to the thionins, and were therefore called γ-thionins (Colilla *et al.*, 1990; Mendez *et al.*, 1990). They have been isolated from seeds of a number of cereals and dicots, including wheat (Colilla *et al.*, 1990), barley (Mendez *et al.*, 1990), sorghum (Bloch and Richardson, 1991), radish (Terras *et al.*, 1992, 1993b, 1995) and potato (Moreno *et al.*, 1994). They inhibit the growth *in vitro* of a wide range of pathogenic fungi, and a protein from radish has been shown to confer resistance to the follar pathogen *Alternaria longipes* when expressed in leaves of transgenic tobacco (see Terras *et al.*, 1995; Broekaert *et al.*, 1995).

73

Ribosome-inactivating proteins have been recently reviewed by Stirpe *et al.*, (1992) and are discussed briefly in section 4 in connection with resistance to viruses. They generally do not inhibit the activity of ribosomes from the plants in which they occur, but show varying degrees of activity against ribosomes of distantly-related species. The RIP from barley seeds has been isolated and shown to inhibit the growth of fungi *in vitro*, in a manner that is synergistically enhanced by enzymes, such as chitinases, that degrade fungal cell walls (Leah *et al.*, 1991). The gene encoding the barley RIP has been transferred into tobacco under the control of a wound-inducible promoter from potato, where it increases the resistance of the plants to soil-borne *Rhizoctonia solani* (Logemann *et al.*, 1992). The protection appears to be mediated by the induced RIP, and there is no evidence that the RIP detrimentally affects the cells of the tobacco plant. The plants were also fertile despite the fact that the wound-inducible promoter is probably active in pollen and flower tissue. The level of resistance was judged to be greater than that conferred upon tobacco plants by a gene for a bacterial chitinase (Jach *et al.*, 1992). The expectation that the simultaneous expression of both RIP and the chitinase will further enhance fungal resistance (Logemann *et al.*, 1992) has recently been confirmed (Jach *et al.*, 1995).

8.4 RESISTANCE BASED ON THE GENETIC MANIPULATION OF SECONDARY METABOLISM

Plants produce a diverse array of low molecular weight, secondary plant products which possess antimicrobial activity. These can be classified either as preformed constitutive inhibitors or as inducible phytoalexin antibiotics. Phytoalexins are antimicrobial compounds that contribute to the plant's active defense against fungal invasion and are synthesised following infection or elicitation. The biochemistry and molecular biology of the phytoalexin response in legumes, such as alfalfa (*Medicago sativa*), is highly advanced. Most legumes produce isoflavonoid-derived phytoalexins that are synthesised by the phenylpropanoid/polymalonate biosynthetic pathway. So far eleven enzymes involved in the formation of pterocarpan phytoalexins have been characterized and cDNAs encoding seven of these enzymes have been isolated (phenylanaline ammonia-lyse, cinnamic acid 4-hydroxylase, 4-coumarate:CoA ligase, chalcone synthase, chalcone reductase, chalcone isomerase and isoflavone reductase, Dixon and Paiva, 1993; Lamb *et al.*, 1992). The key regulatory enzymes of this biosynthetic pathway, phenylalanine ammonia-lyase and chalcone synthase, are known to be encoded by multigene families. Pterocarpan phytoalexins can have a number of different substitution patterns which affect the degree and spectrum of their antifungal activity. These include isoprenyl, methoxy, methylenedioxy and 6a-hydroxy substitutions.

In contrast to legumes, Solanaceous species, such as, tobacco (*Nicotiana tabacum*) and potato (*Solanum tuberosum*), produce sesquiterpene phytoalexins (Bailey and Mansfield, 1982). Unfortunately, far less is known about the enzymology and molecular biology of this group of antimicrobial metabolites. This is also the case for classes of antifungal compounds derived from other plant species. Therefore, the potential to genetically manipulate other groups of inhibitors in a crop species lags behind that for isoflavonoid metabolites in legumes. Nevertheless, a fungal cyclase responsible for the initial folding and cyclization of the building blocks (farnesyl and geranylgeranyl pyrophosphate) for

74

sesquiterpene metabolites has recently been isolated and shown to be expressed in transgenic tobacco (Hohn and Ohlrogge, 1991).

Two main strategies for enhancing the effectiveness of endogenous antifungal metabolites through genetic engineering have been proposed.

i)   Introduction of foreign genes that either lead to expression of novel pathways or to altered structures synthesised by pre-existing pathways.

ii)   Manipulation of genes encoding regulatory enzymes, involved in early committed steps, to alter the flux of precursors through the biosynthetic pathway.

Besides influencing the basic structure and amount of an antifungal metabolite, the ability to genetically alter the process of elicitation of the phytoalexin response could also lead to altered disease resistance. Much research effort is already being directed towards understanding the mechanism of induction of phytoalexin biosynthesis, including identification of elicitors, unravelling the molecular basis of pathogen perception, and elucidating intracellular signal generation processes and transduction pathways (Dixon and Lamb, 1990). Identification of *cis*-elements regulating defense-inducible promoters is of particular value because these could be used to manipulate the levels of expression of foreign genes in order to prevent a build up of potentially toxic metabolites in uninfected healthy cells. However, it should be noted that wound- or infection-inducible genes are invariably regulated in a tissue specific manner. Furthermore, genes encoding an enzyme which utilises or modifies a preexisting metabolite will need to be expressed at the same time and in the same cells as those encoding enzymes of the biosynthetic pathway to be modified (i.e coordinated and cell specific regulated).

8.4.1   Introduction of Foreign Genes

Because different plant species often produce different types of antifungal metabolites, interspecies transfer of biosynthetic or modifying enzymes provides a sound basis for genetically engineering novel types of antifungal metabolite in transgenic plants. Indeed, fungal pathogens are often tolerant of antifungal metabolites produced by their host, but sensitive to structurally-related metabolites from other non-host plants. Resistance to a particular fungal plant pathogen could, therefore, be improved by engineering a plant to produce an antifungal metabolite that is not normally encountered by their pathogens.

There are two possible ways to do this. First, a gene could be introduced which leads to the production of a novel antifungal metabolite not normally found in that plant. This approach has already met with some success in improving disease resistance and involved the introduction of the stilbene synthase gene from grapevines (*Vitis vinifera*) into tobacco (*N. tabacum*) (Hain *et al.,* 1993). Stilbene synthase catalyses the one-step synthesis of the phytoalexin, resveratrol, from *p*-coumaroyl-CoA and malonyl-CoA. Resveratrol is not found in tobacco; however, transgenic tobacco plants containing the grape vine stilbene synthase gene synthesise resveratrol when challenged by fungal infection. In addition, these plants were significantly more resistant to infection by *Botrytis cinerea* than non-transgenic plants. Other opportunities to express foreign phytoalexins in crop plants include the gene encoding casbene synthase (Mau and West, 1994). This gene has been isolated from castor bean (*Ricinus communis*) and encodes an enzyme responsible for the one-step synthesis of the cyclic diterpene phytoalexin, casbene, from geranylgeranyl pyrophosphate. Transgenic plants expressing casbene synthase would, therefore, be expected to exhibit enhanced resistance to fungal pathogens.

The second approach is to improve the fungicidal activity of an existing metabolite by introducing genes which encode enzymes that alter its structure. For example, increasing the lipophilicity of an antifungal metabolite often leads to enhanced activity. In the case of isoflavonoid phytoalexins, the introduction of an isoprenyl moiety is known to increase antifungal activity (Adesanya *et al.*, 1986). This modification could be achieved by expressing prenyltransferase genes from other organisms. A suitable prenyltransferase has already been purified from bean (*Phaseolus vulgaris*) and its gene could be cloned for this purpose (Biggs *et al.*, 1987). Methylation of isoflavonoid phytoalexins is also known to enhance antimicrobial activity. In this case, a number of methyltransferases of isoflavonoid phytoalexins have been isolated and their genes are being isolated for expression in transgenic plants. An alternative way to increase the effectiveness of an antifungal metabolite is to alter the stereochemistry of an already active metabolite using genes encoding enzymes, such as epimerases, that introduce chiral centres. In the case of isoflavonoid phytoalexins, manipulation of the stereochemistry or the stereospecific interconversion of active metabolites during biosynthesis could increase production of the more effective isomeric components which are less susceptible to detoxification by fungal pathogens (Dixon and Paiva, 1993). Finally, there is also scope for introducing genes that synthesise antifungal metabolites from other organisms. Transgenic expression of polyketides antibiotics is particularly attractive because these metabolites are known to be synthesised by a single multifunctional protein in actinomycetes and fungi (Hopwood and Sherman, 1990).

8.4.2   Manipulation of Genes Encoding Regulatory Enzymes

Genetic manipulation of the amount or the rate of production of an antifungal metabolite provides another approach to improving crop resistance to fungal pathogens. Increasing the substrate supply for either pre-existing or novel metabolites could lead to enhanced levels of the antifungal metabolite and, thereby, increased efficacy against fungal pathogens. Potential target enzymes for increased expression include acetyl-CoA carboxylase, hydroxy methylglutaryl-CoA reductase and phenylalanine ammonia-lyase. These enzymes are encoded by multigene families and increases in the activities of specific metabolites could be achieved by placing these genes under the control of a strong constitutive promoter or by increasing the copy number of the genes. Alternatively, modulation of the activities of defense-related promoters could allow the amounts and the level of synthesis of specific antifungal metabolites to be altered.

In conclusion, there is now significant information on the biochemistry and molecular biology of phytoalexin biosynthesis, particularly in legumes, to attempt to genetically manipulate antifungal metabolite production in plants. Nevertheless, more information, particularly on fundamental aspects of enzymology and molecular biology of secondary product biosynthesis, is still needed. This includes information on the type and source of substrates for biosynthesis, the key regulatory or rate-limiting steps and branch points, and the rates of turn over of the metabolites within the cell. In addition, further information on the quantitative structure-antifungal activity relationships of analogues to be expressed in transgenic plants is required, so that promising metabolites with increased activity against target pathogens can be identified. Furthermore, suitable sources of genes that encode enzymes which modify the structures of chosen metabolites or their precursors will also need to be identified.

8.5   CONCLUSIONS

The use of the techniques of genetic manipulation to investigate the mechanisms by which plants successfully resist fungal pathogens is now a very active field of research. Major advances include the isolation and characterisation of some R genes. Plants have been made resistant to fungi by transforming them to produce antifungal enzymes and toxins. The most intensively studied of these antifungal proteins are the chitinases, and it is clear that to produce useful commercial crops they will need to be used in combination with other enzymes or toxins with which they act synergistically. There is, internationally, an active search to identify from plant sources other antifungal proteins that could be similarly used.

Secondary metabolites of plants, including phytoalexins, provide a rich source of antifungal compounds, whose activity against specific pathogens could be enhanced by genetic manipulation. Thus, transformation of tobacco to produce the phytoalexin reservatrol, increased resistance to *B. cinerea*. Other opportunities to express "foreign" phytoalexins are currently possible. However the full exploitation of this method of enhancing natural resistance requires more basic information on the enzymology and molecular biology of the biosynthesis of secondary metabolites.

8.6   REFERENCES

Adesanya, S.A., O'Neil, M.J. & Roberts, M.F. (1986). Structure-related fungi toxicity of isoflavonoids. *Physiological and molecular plant pathology*, **29**, 95-103.

Alexander, D., Goodman, R.M., Gut-Rella, M. *et al.*, (1993). Increased tolerance to two oomycete pathogens in transgenic tobacco expressing pathogenesis-related protein 1a. *Proceedings of the National Academy of Sciences of the USA* **90**, 7327-7331.

Bailey, J.A. & Mansfield, J.A. (1982) Phytoalexins. Blackie, Glasgow

Bennetzen J.L. & Jones J.D.G. (1992). Approaches and progress in the molecular cloning of plant disease resistance genes. In *Genetic Engineering, Principles and Methods* (K.Setlow ed.), Vol 14, pp 99-124. Plenum Press, New York.

Bent A.F., Kunkel B.N., Dahlbeck D. *et al.*, (1994). *RPS2* of *Arabidopsis thaliana*: a leucine-rich repeat class of plant disease resistance gene. *Science*, **265**, 1856-1859.

Biggs, D.R.M., Welle, R., Visser, F.R. & Grisebach, H. (1987) Dimethylallylpyrophosphate: 3,9-dihydroxypterocarpan 10-dimethylallyl transferase from *Phaseolus vulgaris*. *FEBS letters*, **220**, 223-226.

Bloch, C. & Richardson, M. (1991). A new family of small (5 kDa) protein inhibitors of insect α-amylases from seeds of sorghum (*Sorghum bicolor* (L) Moench) have sequence homologies with wheat γ-purothionins. *FEBS Letters*, **279**, 101-104.

Bol, J.F., Linthorst, H.J.M. & Cornelissen, B.J.C.(1990). Plant pathogensis-related proteins induced by virus infection. *Annual Review of Phytopathology*, **28**, 113-138.

Broekaert, W.F., Parijs, J.V., Leyns, F. *et al.*, (1989). A chitin-binding lectin from stinging nettle rhizomes with antifungal properties. *Science*, **245**, 1100-1102.

Broekaert, W.F., Marien, W. & Terras, F.R.G. (1992). Antimicrobial peptides from Amaranthus caudatus, seeds with sequence homology to the cysteine/glycine-rich domain of chitin-binding proteins. *Biochemistry*, 31, 4308-4314.

Broekaert, W.F., Terras, F.R.G., Cammue, B.P.A. and Osborn, R.W. (1995). Plant defensins: novel antimicrobial peptides as components of the host defense system. *Plant Physiology*, **108**, 1353-1358.

Broglie, K., Chet, I., Holliday, M. *et al.*, (1991). Transgenic plants with enhanced resistance to the fungal pathogen *Rhizoctonia solani*. *Science*, **254**, 1194-1197.

Broglie, R. & Broglie, K. (1993). Chitinase gene expression in transgenic plants: a molecular approach to understanding plant defence responses. *Philosophical Transactions of the Royal Society, Series B London*, **342**, 265-270.

Cammue, B.P.A., De Bolle, M.F.C., Terras, F.R.G. *et al.*, (1992). Isolation and characterization of a novel class of antimicrobial peptides from *Mirabilis jalapa* seeds. *Journal of Biological Chemistry*, **267**, 2228-2233.

Cammue, B.P.A., Thevissen, K., Hendriks, M. *et al.*, (1995). A potent antimicrobial protein from onion seeds showing sequence homology to plant lipid transfer proteins. *Plant Physiology*, **109**, 445-455.

Carmona, M.J., Molina, A., Fernandez, J.A. *et al.*, (1993). Expression of the α-thionin gene from barley in tobacco confers enhanced resistance to bacterial pathogens. The *Plant Journal*, 3, 457-462.

Colilla, F.J., Rocher, A. & Mendez, E. (1990). Gamma-purothionins: amino acid sequence of two polypeptides of a new family of thionins from wheat endosperm. *FEBS Letters*, **270**, 191-194.

Collinge, D.B., Kragh, K.M., Mikkelsen, J.D. *et al.*, (1993). Plant chitinases. *The Plant Journal*, 3, 31-40.

Collinge, D.B., Gregersen, P.L. & Thordal-Christensen, H. (1994). The induction of gene expression in response to pathogenic microbes. In *Mechanisms of Plant Growth and Improved Productivity: Modern Approaches and Perspectives* (A.S.Basra ed.) pp. 391-433. Marcel Decker: New York.

Cornelissen, B.J.C., Hooft van Huijsduijen, & Bol, J.F. (1986). A tobacco mosaic virus-induced tobacco protein is homologous to the sweet-tasting protein thaumatin. *Nature*, **321**, 531-532.

Cornelissen B.J.C. & Melchers L.S. (1993). Strategies for control of fungal diseases with trangenic plants. *Plant Physiology*, 101, 709-712.

De Bolle, M.F.C., Eggermont, K., Duncan, R.E. *et al.*, (1995). Cloning and characterization of two cDNA clones encoding seed-specific antimicrobial peptides from *Mirabilis jalapa* L. *Plant Molecular Biology*, **28**, 713-721.

de Wit, P.J.G.M. (1992). Molecular characterization of gene-for gene systems in plant fungus interactions and the application of avirulence genes in control of plant pathogens. *Annual Review of Phytopathology*, **30**, 391-418.

Dixon, R.A. & Lamb, C.J. (1990) Molecular communication in interactions between plants and microbial pathogens. *Annual Review of Plant Physiology and Plant Molecular Biology*, **41**, 339-367.

Dixon, R.A. & Paiva, N.L. (1993) Prospects for the genetic manipulation of antimicrobial plant secondary products. BCPC monograph No. 55: Opportunities for molecular biology in crop protection, 113-119.

Florack, D.E.A. & Stiekema, W.J. (1994). Thionins: properties, possible biological roles and mechanisms of action. *Plant Molecular Biology*, **26**. 25-37.

Hain, R., Reif, H-J., Krause, E., Langbartels, R., Kindl, H., Vornam, B., Wiese, W., Schmelzer, E., Schreier, P.H., Stocker, R.H. & Stenzel, K. (1993) Disease resistance results from foreign phytoalexin expression in a novel plant. *Nature*, **361**, 153-156.

Hammond-Kosack K.E., Harrison K. & Jones J.D.G. (1994a). Developmentally regulated cell death on expression of the fungal avirulence gene *Avr9* in tomato seedlings carrying the disease-resistance gene Cf-9. *Proceedings of the National Academy of Science of the USA*, **91**, 10445-10449.

Hammond-Kosack K.E., Jones D.A. & Jones J.D.G. (1994b). Identification of two genes required in tomato for full *Cf-9*-dependent resistance to *Cladosporium fulvum*. *The Plant Cell*, **6**, 361-374.

Hejgaard, J., Jacobsen, S. & Svendsen, I (1991). Two antifungal thaumatin-like proteins from barley grain. *FEBS Letters*, **291**, 127-131.

Hohn, T.M. & Ohlrogge, J.B. (1991). Expression of a fungal sesquiterpene cyclase gene in transgenic tobacco. *Plant physiology*, **97**, 460-462.

Hopwood, D.A. & Sherman, D.H. (1990). Molecular genetics of polyketides and its comparison to fatty acid biosynthesis. *Annual Review of Genetics*, **24**, 37-66.

Huynh, Q.K., Borgmeyer, J.R. & Zobel, J.F. (1992). Isolation and characterization of a 22 kDa protein with antifungal properties from maize seeds. *Biochem. Biophys. Res. Communic.* **182**, 1-5.

Jach, G., Logemann, S., Wolf, G. *et al.*, (1992). Expression of a bacterial chitinase leads to improved resistance of transgenic tobacco plants against fungal infection. *Biopractice* **1**, 33-40.

Jach, G., Gornhardt, B., Mundy, J. *et al.*, (1995). Enhanced quantitative resistance against fungal disease by combinatorial expression of different barley antifungal proteins in transgenic tobacco. *Plant Journal*, **8**, 97-109.

Johal, G.S. & Briggs S.P. (1992). Reductase activity encoded by the *HM1* disease resistance gene in maize. *Science*, **258**, 985-987.

Jones, D.A., Thomas C.M., Hammond-Kosack, K.E. *et al.*, (1994). Isolation of the tomato *Cf-9* gene for resistance to *Cladosporium fulvum* by transposon tagging. *Science*, 266, 789-793.

Knogge, W. (1994). Plant disease resistance genes-7th International Symposium on Molecular Plant-Microbe Interactions. *European Journal of Plant Pathology*, **100**, 283-286.

Lamb, C.J., Ryals, J.A., Ward, E.R. & Dixon, R.A. (1992). Emerging strategies for enhancing crop resistance to microbial pathogens. *Bio/Technology*, **10**, 1436-1445.

Leah, R., Tommerup, H. Svendsen, I. *et al.*, (1991). Biochemical and molecular characterization of three barley seed proteins with antifungal properties. *Journal of Biological Chemistry*, **266**, 1564-1573.

Linthorst, H.J.M., Meuwissen, R.L.J., Kauffman, S. *et al.*, (1989).Constitutive expression of pathogenesis-related proteins PR-1, GRP and PR-S in tobacco has no effect on virus infection. *The Plant Cell*, **1**, 285-291.

Liu, D., Raghothama, K. G., Hasegawa, P. M. *et al.*, (1994). Osmotin over expression in potato delays development of disease symptoms. *Proceedings of the National Academy of Sciences of the USA*, **91**, 1888-1892.

Logemann, J., Jach, G., Tommerup, H. *et al.*, (1992). Expression of a barley ribosome-inactivating protein leads to increased fungal protection in transgenic tobacco plants. *Bio/Technology*, **10**, 305-308.

Martin, G.B., Brommonschenkel, S.H., Chungwongse, J. *et al.*, (1993). Map-based cloning of a protein kinase gene conferring disease resistance in tomato. *Science*, **262**, 1432-1436

Mauch, F., Mauch-Mani, B. & Boller, T. (1988). Antifungal hydrolases in pea tissue. *Plant Physiology*, **88**, 936-942.

Mau, C.J.D. & West, C.A. (1994). Cloning of casbene synthase cDNA-evidence for concerned structural features among terpenoid cyclases in plants. *Proceedings of the National Academy of Sciences of the USA*, **91**, 8497-8501.

Melchers, L.S., Sela-Buurlage, M.B., Vloemans, S.A. *et al.*, (1993). Extracellular targeting of the vacuolar proteins AP24, chitinase and β-1:3-glucanase in transgenic plants. *Plant Molecular Biology*, **21**, 583-593.

Melchers, L.S., Apotheker de Groot, M., van der Knaap, J.A. *et al.*, (1994). A new class of tobacco chitinases homologous to bacterial exo-chitinases displays antifungal activity. *The Plant Journal*, **5**, 469-480.

Mendez, E., Moreno, A., Colilla, F. *et al.*, (1990). Primary structure and inhibition of protein synthesis in eukaryotic cell-free system of a novel thionin, γ-hordothionin, from barley endosperm. *European Journal of Biochemistry*, **194**, 533-539

Mindrinos, M,, Katagiri, F., Yu G-L. *et al.*, (1994). The *A. thaliana* disease resistance gene *RPS2* encodes a protein containing a nucleotide-binding site and leucine-rich repeats. *Cell*, **78**, 1089-1099.

Moffat A.S. (1994) Mapping the sequence of disease resistance. *Science*, **265**, 1804-1805.

Molina, A. & Garcia-Olmedo, F (1993). Developmental and pathogen-induced expression of three barley genes encoding lipid transfer proteins. *The Plant Journal*, **4**, 983-991.

Moreno, M., Segura, A. & Garcia-Olmedo, F. (1994). Pseudothionin, a potato peptide active against potato pathogens. *European Journal of Biochemistry*, **223**, 135-139.

Neuhaus, J-M., Ahl-Goy, P., Hinz, U. *et al.*, (1991). High-level expression of a tobacco chitinase gene in *Nicotiana sylvestris*. Susceptibility of transgenic plants to *Cercospora nicotianae* infection. *Plant Molecular Biology*, **16**, 141-151.

Neuhaus, J-M., Flores, S., Keefe, D. *et al.*, (1992). The function of vacuolar β-1:3 glucanase investigated by antisence transformation. Susceptibility of transgenic *Nicotiana sylvestris* plants to *Cercospora nicotianae* infection. *Plant Molecular Biology*, **19**, 803-813.

Newbury H.J. (1992). Fungal resistance; the isolation of a plant R gene by transposon tagging. In *Plant Genetic Manipulation for Crop Protection* (A.M.R. Gatehouse, V.A. Hilder and D. Boulter eds), pp 109-134. C.A.B. International. Wallingford: UK

Niderman, T., Genetet, I., Bruyère, T. *et al.*, (1995). Pathogenesis-related PR-1 proteins are antifungal. Isolation and characterization of three 14-kilodalton proteins of tomato and of a basic PR-1 of tobacco with inhibitory activity against *Phytophthora infestans*. *Plant Physiology*, **108**, 17-27

Parijs, J.V., Broekaert, W.F., Goldstein, I.J. *et al.*, (1991). Hevein: an antifungal protein from rubber-tree (*Hevea brasiliensis*) latex. *Planta*, **183**, 258-264.

Payne, G., Ward, E., Gaffney, T. *et al.*, (1990). Evidence for a third structural class of β-1:3-glucanase in tobacco. *Plant Molecular Biology*, **15**, 797-808.

Pierpoint, W.S., Tatham, A.S. & Pappin, D.J.C. (1987). Identification of the virus-induced protein of tobacco leaves that resembles the sweet protein thaumatin. *Physiological Molecular Plant Pathology*, **31**, 291-298.

Ponstein, A.S., Bres-Vloemans, S.A., Sela-Buurlage, M.B. *et al.*, (1994). A novel pathogen and wound inducible tobacco (*Nicotiana tabacum*) protein with antifungal activity. *Plant Physiology*, **104**, 109-118.

Sela-Buurlage, M.B., Ponstein, A.S., Bres-Vloemans, S.A. *et al.*, (1993). Only specific tobacco (*Nicotiana tabacum*) chitinases and β-1:3-glucanases exhibit antifungal activity. *Plant Physiology*, **101**, 857-863.

Stirpe, F., Barbieri, L., Battelli, M.G. *et al.*, (1992). Ribosome-inactivating proteins from plants: present status and future prospects. *Bio/Technology*, **10**, 405-412.

Terras, F.R.G., Schoofs, H.M.E., De Bolle, M.F.C. *et al.*, (1992). Analysis of two novel classes of plant antifungal proteins from radish (*Raphanus sativus*) seeds. *Journal of Biological Chemistry*, **267**, 15301-15309.

Terras, F.R.G., Schoofs, H.M.E., Thevissen, K. *et al.*, (1993a). Synergistic enhancement of the antifungal activity of wheat and barley thionins by radish and oilseed rape 2S albumins and barley trypsin inhibitors. *Plant Physiology*, **103**, 1311-1319.

Terras, F.R.G., Torrekens, S., Van Leuven, F. *et al.*, (1993b). A new family of basic cysteine-rich plant antifungal proteins from Brassicaceae-species. *FEBS Letters*, **316**, 233-240.

Terras, F.R.G., Eggermont, K., Kovaleva, V. *et al.*, (1995). Small cysteine-rich antifungal proteins from radish: their role in host defense. *The Plant Cell*, **7**, 573-588.

Thoma, S., Kaneko, Y. & Somerville, C. (1993). A non-specific lipid transfer protein from *Arabidopsis* is a cell wall protein. *The Plant Journal*, **3**, 427-436.

Toubart, P., Desiderio, A., Salvi, G. *et al.*, (1992). Cloning and characterization of the gene encoding the endopolygalacturonase-inhibiting protein (PGIP) of *Phaseolus vulgaris*. *The Plant Journal*, **2**, 367-373.

van den Elzen, P.J.M., Jongedijk, E., Melchers, L.S. *et al.*, (1993). Virus and fungal resistance: from laboratory to field. *Philosophical Transactions of the Royal Society. Series B London*, **342**, 271-278

van Loon, L.C. (1985). Pathogenesis-related proteins. *Plant Molecular Biology*, **4**, 111-116.

Vigers, A.J., Roberts, W.K. & Selitrennikoff, C.P. (1991). A new family of plant antifungal proteins. *Molecular Plant-Microbe Interactions*, **4**, 315-323.

Vigers, A.J., Wiedemann, S., Roberts, W.K. *et al.*, (1992). Thaumatin-like pathogenesis proteins are antifungal. *Plant Science*, **83**, 155-161.

Whitham, S., Dinesh-Kumar, S.P., Choi, D. *et al.*, (1994). The product of the tobacco mosaic virus resistance gene N; similarity to Toll and the Interleukin-1 receptor. *Cell*, **78**, 1101-1115.

Woloshuk, C.P., Meulenhoff, J.S., Sela-Buurlage, M. *et al.*, (1991). Pathogen-induced proteins with inhibitory activity towards *Phytophthora infestans*. *The Plant Cell*, 3, 619-628.

Zhu, Q., Maher, E.A., Masoud, S. *et al.*, (1994). Enhanced protection against fungal attack by constitutive co-expression of chitinase and glucanase genes in transgenic tobacco. *Bio/Technology*, **12**, 807-812.

9.   MODIFYING   RESISTANCE   TO   PLANT-PATHOGENIC BACTERIA

D.E. Stead and J.G. Elphinstone
(MAFF Central Science Laboratory)

9.1   BACKGROUND

9.1.1   Economic Importance of Plant Pathogenic Bacteria in North-West European Agriculture.

Pathogenicity for higher plants has evolved independently many times in the Prokaryotes. Plant pathogens are found in the Mollicutes (spiroplasmas, mycoplasma and mycoplasma-like organisms), the Firmicutes (Gram positive bacteria) and Gracilicutes (Gram negative bacteria). In the latter group plant pathogens occur in all the major sub-classes.

Bacteria are responsible for some very destructive plant diseases. Although perhaps causing greater crop loss in the tropics, there are several important temperate diseases which either occur in, or currently threaten UK and North European crops. There are few chemicals available for controlling bacterial diseases and little commercial prospect of their development or approval as pesticides largely because of social, political and economic constraints.

Perhaps the greatest economic losses in Europe are caused by subspecies of *Erwinia carotovora* which cause soft rot of a wide range of vegetables. Potatoes are particularly badly affected both in the field (blackleg caused by *E. carotovora* subsp. *atroseptica*) and in store (rotting caused by *E. carotovora* subsp. *atroseptica* and subsp. *carotovora*). Together with field and post harvest soft rot of other vegetables, they probably cause £25-£50 million crop loss per year. Three other bacterial pathogens currently threaten the European potato crop although they are not yet established in the UK largely through certification and other statutory requirements for their absence from imported plant material. These include potato ring rot (*Clavibacter michiganensis* subsp. *sepedonicus*), brown rot (*Pseudomonas solanacearum*) and slow wilt (*Erwinia chrysanthemi*). It is assumed that ring rot alone if allowed to become established in the UK, could result in a £20 million per year loss to the industry.

Other economically important diseases in Europe include bacterial blight of pea (*Pseudomonas syringae* pv. *pisi*), bacterial canker of tomato (*Clavibacter michiganensis* subsp. *michiganensis*), fireblight of pear, apple and ornamental Pomoideae, and Pelargonium blight (*Xanthomonas campestris* pv. *pelargonii*).

9.1.2   Importance and Current understanding of the development of Resistance to Bacterial Pathogens in Crops

Development of resistant plants has for sometime been seen as the best means for control but there have been relatively few successes largely because the mechanisms of pathogenesis and resistance remain poorly understood at the molecular level.

Conventional breeding for resistance has centred on crossing susceptible commercial cultivars with naturally resistant cultivars or closely related species. Prospects for breeding resistance to bacteria in potatoes are reviewed by Elphinstone (1994). The work of Taylor and his colleagues on bacterial halo blight of beans and blight of peas caused by *Pseudomonas syringae* pv. *phaseolicola* and *P. syringae* pv. *pisi* respectively, has led to a much better understanding of the genetics of resistance to bacterial pathogens. Each plant species may have up to five different resistance genes. In any one cultivar any number of these may be present. Resistance in these cases is based on the gene-for-gene relationship in which, if a plant resistance (R) gene recognises a pathogen's avirulence (avr) gene, the plant defends itself by producing low molecular weight antibiotic compounds (phytoalexins) followed by other compounds including proteins or peptides which together result in rapid death of the plant cells around the area of infection. This process is known as the hypersensitive reaction. If a particular cultivar lacks the R gene corresponding to the pathogen's avr gene, it will succumb to infection and subsequent disease. Because there are often several different R genes involved in resistance and hence a corresponding number of avr genes available, many pathogens have developed into a series of 'races' which differ in the number of avr genes they possess. Hence *Xanthomonas campestris* pv. *malvacearum* causing blight of cotton has some 20 races and *P. syringae* pv. *pisi* has 7 races. (Taylor *et al.*, 1989).

Whereas much recent research on bacterial resistance has centred on R and avr genes and their chemical nature and action, only very recently has success with engineered resistance based on these mechanisms been achieved (Martin *et al.*, 1993, Shields & Stratford, 1993). It is to be expected that many other developments will soon be made. However, there are other ways in which resistance may be modified or introduced and these methods do not necessarily use defence mechanisms derived from plants.

9.2   STRATEGIES FOR MANIPULATION OF BACTERIAL RESISTANCE

Resistance genes can be introduced into plants either by breeding, ie crossing with plant species or cultivars with known resistance, or by introducing selected genes into plant chromosomes using genetic engineering techniques.

As is the case for the transfer of many genes into plant species, transformation with resistance genes to bacteria has mainly involved the use of the bacterial genus *Agrobacterium*, using either tumorigenic or rhizogenic (Ti or Ri) vector plasmids which have been disarmed in an attempt to prevent them becoming pathogenic in their own right. Some of the practical problems are briefly discussed in section 2.5. Other bacterial plant pathogens also have potential as vectors including *Rhodococcus fascians* and *Clavibacter xyli* subsp. *cynodontis* (Turner *et al.*, 1993) although the mechanisms by which they introduce genes are less well documented.

9.2.1   Mapping, Isolation and Manipulation of Natural Plant Vertical Resistance (R) Genes

The most obvious way of introducing resistance is undoubtedly through finding genes for resistance to specific pathogens and introducing them into the plant genomes from which they are absent. This has recently been achieved in tomato resulting in increased resistance to *Pseudomonas syringae* pv. *tomato* causing bacterial speck disease (Martin *et al.*, 1993; Salmeron *et al.*, 1994). The theory behind such strategies is well reviewed by Shields and Stratford (1993) and Chasan (1994). Having identified the avr Pto gene in the pathogen, the Pto resistance gene was located. Susceptible tomato plants transformed with complementary DNA corresponding to a mRNA transcript from this region became resistant to races of pathogen carrying avr Pto, but not to other races of pathogen. The gene product was then found to be a serine-threonine protein kinase. This result was unexpected and illustrates the important point that although R and avr genes are now being identified(see also section 8), we still do not know how their gene products react with one another leading to resistance, although we know that it often involves the hypersensitive reaction. One other important discovery is that some avr genes are common to several related pathogens.

Because the hrp regions of the bacterial chromosomes responsible for pathogenesis and including the avr gene clusters are sometimes common between unrelated bacteria, it is thought that genetic exchange between such bacteria has occurred. When transferred from *P. syringae* pv. *tomato* to *P. syringae* pv. *glycinae* (a soybean pathogen), the avr Pto and avr D genes made some strains avirulent towards soybeans implying a race specific association (Ronald *et al.*, 1992). Similarly a resistance locus in the crucifer *Arabidopsis* has been identified using avr genes from *Pseudomonas syringae* pv. *glycinea*, a soybean pathogen (Innes *et al.*, 1993). Because avirulence genes from bacteria specific for different plants can appear to be identical, it is possible that R genes cloned from one plant species may function in another (Dangl *et al.*, 1992). Although this type of research may well result in plants with resistance to damaging diseases there are some practical problems. All avr genes from a particular pathogen must be recognised and the requisite R genes introduced, otherwise resistance will only be to specific races of a pathogen. Race distribution within populations can change rapidly as it did with *P. syringae* pv. *pisi*, which in Europe was largely an occasional pathogen of vining peas. The fairly recent introduction of closely related "combining" peas for protein production, derived from different genetic stock to vining peas and hence with different R genes, resulted in a rapid change in the distribution of races of pv. *pisi*. In particular races, 2, 4 and 6 rapidly spread throughout Europe including the UK (Stead, unpublished). Niches would rapidly be created for uncommon races to become more widespread through control of other races. Gene-for-gene resistance is also not always particularly durable and expensively produced resistant cultivars could thus have short lives. Also most plant resistance does not seem to be of the gene-for-gene type. Nevertheless further understanding of the avr and R gene products and the ways in which they interact and cause hypersensitivity will undoubtedly increase the chances of successful engineering of resistance in specific cases (see also section 8).

9.2.2 Selection, Incorporation and Expression of Anti-toxin Genes

Many bacteria, including the pathovars of *Pseudomonas syringae* produce toxins which play an important but often unknown part in disease. Many of the toxins which have been characterised are from pathovars of *Pseudomonas syringae*, and are involved in the production of chlorosis, often seen as a halo of yellow tissue around the necrotic lesion as in halo blight of bean caused by *P. syringae* pv. *phaseolicola* (*Psp*) which produces phaseolotoxin. Other toxins include syringomycin and tabtoxin. These toxins are often also anti-bacterial, presumably to reduce competition from other bacteria. Obviously there must be some mechanism by which bacteria are protected from the toxins they produce. *Pseudomonas syringae* pv. *phaseolicola* produces two ornithine carbamoyltransferase enzymes (OCTases) which are known to be involved in the metabolism of arginine and glutamate. Phaseolotoxin, which is specific for disease in *Phaseolus vulgaris* is produced by *Psp* and inhibits some OCTases. One of the *Psp* OCTases is sensitive to the toxin and the other is resistant. Several groups of workers have been involved in isolating the gene for the phaseolotoxin resistant OCTase (Hatziloukas and Panapoulos, 1989; Fuente *et al.*, 1993). Transgenic plants expressing this toxin-resistant OCTase were no longer susceptible to the effect of phaseolotoxin, and in fact, showed a hypersensitive response when challenged with the pathogen.

Similar work on the tabtoxin of the wildfire bacterium *P. syringae* pv. *tabaci* (Yoneyama and Anzai, 1993) has shown that transfer of tabtoxin resistance genes, presumably responsible for an OCTase production, increased resistance of tobacco to wildfire disease.

9.2.3   Transfer and Expression of Genes Encoding other Plant Resistance Factors

Various plant products and secondary metabolites are linked to resistance, and genes for their production could be used to induce novel resistance. These include phytoalexins, which are antibiotic chemicals produced in response to the presence of the pathogen, and compounds such as lignin which act as physical barriers to prevent pathogen spread. Although such possibilities are discussed at conferences, there is little documented work.

There is more current interest in antibacterial proteins. Potato chitinases and β 1-3 glucanases (Laflamme & Roxby, 1989) which may be expected to have anti-fungal activity, also appear to have activity against the ring rot bacterium, *Clavibacter michiganensis* subsp. *sepedonicus*. Potatoes could be engineered to produce such enzymes with greater efficiency.

Although plants produce lysozymes, their role in the lysis of bacteria is not understood, and non-plant lysozymes are more effective. These are discussed in section 9.2.4.

Thionins. Thionins are cysteine-rich, 5 kDa polypeptides present in various cereals which are toxic to a wide range of plant pathogenic bacteria (Carmona *et al.*,1993, Florack *et al.*, 1993) : (see also section 8). Those isolated from the endosperm of barley are termed hordothionins and those from wheat are termed purothionins. They are highly basic. Being fairly small molecules they have been sequenced and synthetic genes for them produced and introduced into tomato and potato (Dons *et al.*,1991). Although there is good evidence of *in vitro* activity against some *Pseudomonas syringae* pathovars, *Clavibacter michiganensis* subspecies and *Xanthomonas campestris* pathovars, primarily those from tobacco, tomato and potato, there is only a single example of successful disease control. This was in transgenic tobacco with a α-hordothionin gene which produced a hordothionin with identical properties to the wild type protein, and which had good *in vitro* antibacterial activity. Transgenic plants showed increased resistance to two *Pseudomonas syringae* pathovars (pv. *syringae* and pv. *tabaci*).

9.2.4   Introduction and Expression of Non-Plant Resistance Genes

Several proteins or peptides are produced by organisms other than plants, in defence against bacteria and other micro-organisms. These include lysozymes from various sources, lytic peptides produced by some invertebrates and antibodies produced by vertebrates.

Lysozymes. Although lysozymes are found in plants little is known about them, and their suggested role in defence against bacteria has yet to be proven. Some of these plant lysozymes are multifunctional, for example, they also break down chitin. However, their ability to break peptidoglycan in bacterial cell walls is much less than that of some mammalian and phage lysozymes.

The bacteriophage T4 lysozyme is the most active anti-bacterial lysozyme known, and is effective against Gram positive as well as Gram negative bacteria (During *et al.*, 1993). It has been shown that the T4 phage normally infecting *E. coli* can also replicate as a lytic phage in some *Erwinia carotovora* strains (Pirhonen & Palva, 1988). Since *E. carotovora* is active in the intercellular spaces of plant tissues such as potato tubers, the phage lysozyme gene was coupled to a sequence encoding the α-amylase signal peptide which would lead to secretion from the cell and hence localisation of the lysozyme in the intercellular spaces of transgenic potato plants (During *et al.*, 1993). During and co-workers then challenged the transgenic plants to *E. carotovora* infection and showed that tissue maceration was significantly reduced. Similar work involving the incorporation of genes for chicken egg white lysozyme into transgenic tobacco plants was discussed by Destefano-Beltran *et al.*, (1993) and Trudel *et al.*, (1992). Research at the International Potato Center will concentrate on the transfer into potato of gene constructs for bovine lysozyme which has a better stability in potato tissues than the chicken egg-white lysozyme (M. Ghislane, pers. comm.). A general review on the use of lysozymes is given by During (1993).

Lytic peptides. Some animals (insects, molluscs, amphibians, mammals) produce small lytic peptides that are involved in their defence against micro-organisms including bacteria. These include cecropins, maganins, attacin and melittin, which have some activity against nematodes and fungi as well as bacteria, but are not phytotoxic. Cecropins, first found in the giant silk moth, are the most studied group. They are 35 amino acids in length and contain hydrophobic and hydrophilic areas. The gene for Cecropin B has been cloned and the peptide synthesised. The gene has been successfully introduced into several plant species. However several synthetic cecropins which have improved characteristics have also been developed. One of these, called Shiva-1, has 46% homology to the wild type Cecropin B and excellent general antibacterial activity *in vitro*. The gene for Shiva-1 was chemically synthesised and inserted into tobacco plants. Progeny plants showed marked resistance to a highly virulent *Pseudomonas solanacearum* strain (Jaynes *et al.*, 1993). Symptoms were delayed, and disease severity and mortality were reduced (Jia, 1993). Nascari *et al.*, (1991), Montanelli & Nascari (1991) and Watanabe *et al.*, (1993) described similar results for cecropin-producing transgenic potatoes challenged also with *P. solanacearum*. However, despite early success in obtaining transgenic plants showing transcription of introduced gene constructs, a lack of increased resistance to *Erwinia caratovora* was concluded to be due to instability of the peptide in potato tissues (Allefs *et al.*, 1995). Further work involving transfer of gene constructs for other antibacterial peptides, showed that resistance to *E. caratovora* in potato cultivars was slightly increased in transgenic potatoes which expressed constructs encoding tachyplesin I but was not affected in those expressing α-hordothionin (Allefs,1995). These results reflected the *in vitro* toxicities of the two peptides to *Erwinia*.

It is also reported (EPPO, 1994) that an American group have introduced a gene for the lytic peptide Attacin E into M26 apple rootstocks. These transgenic plants were more resistant to the fireblight bacterium, *Erwinia amylovora*. Work with scion varieties of apples, e.g. Gala, continues.

Plantibodies. Animal genes encoding antibodies to several bacteria, viruses and fungi have been successfully expressed in plants (Hiatt *et al.*,1989). Since bacteria have a very wide range of epitopes for antibody production, many of which are common to non-pathogens, selection of the epitope is critical for successful development of resistance. Specific enzymes produced by pathogens which are essential for pathogenicity are good targets for plantibodies. It is necessary, however, to consider that the plantibody must be transported to the site of infection on the surface of cells so that the pathogenic enzyme can be inactivated before it can carry out its function. Antibodies to host pathogen-receptor molecules also have the potential to induce resistance.

Antisense genes. There is one record (Walter, 1991) of an attempt to engineer resistance in grapevine to pathogenic agrobacteria, *A. tumefaciens* and *A. vitis*, by introducing *A. tumefaciens* oncogenes in an antisense form. The results are not recorded.

9.2.5   Gene transfer with *Agrobacterium* spp

One basic assumption when using plant pathogenic bacteria such as *Agrobacterium* spp. to introduce genes into plant tissue is that once the plant genome has been transformed, the vector bacterium can be eradicated. This is necessary not only to prevent disease (although such strains have usually been disarmed) but also to prevent the binary vector meeting wild type agrobacteria and transferring genetic information thus allowing gene escape. Recent but as yet unpublished work indicates that conventional antibiotics often fail to eradicate *Agrobacterium spp* from genetically engineered plants, and that the binary vector itself can survive for many months. Although the bacterium is unlikely to be transmitted through seed from one generation of plants to the next, this is an aspect of genetic modification that must be studied more closely and the risks eliminated if such modifications are to be acceptable. Vector elimination should not simply be judged by lack of tumorigenic or rhizogenic activity, but tested for by using appropriate genetic probes. Little epidemiological work on agrobacteria, their latency and disease expression has been done as they are not often of economic importance, and are usually only seen as causing quality problems in ornamental plants.

9.3 DISCUSSION

The development of the R gene- and avr gene-engineered resistance has occurred as a result of painstaking research on the genetics of the host-pathogen interaction. Further development in this area could be limited unless more is known of the gene products and the way they interact to cause or prevent the hypersensitive reaction. Likewise, it will be difficult to incorporate genetic resistance mechanisms other than those involved in the gene-for-gene mechanism, unless more is known about the way they work in the plant. Hence there is currently much research on the types and rates of accumulation of pathogenesis-related proteins, peroxidases and hydrolytic enzymes in resistant and susceptible cultivars. Implicit in this research is the elucidation of the regulatory factors involved. One potential "growth" area is in the investigation of the sources of active oxygen production and membrane-bound oxidase activity which appear to be involved in the response of the plant to pathogens.

Another way to identify the essential components of resistance is to produce susceptible mutants in plants which are resistant to a specific pathogen (Salmeron et al.,1994). However, it must be recognised that success will be most likely in plants with smaller genomes such as *Arabidopsis* (Dangl et al., 1992; Dangl et al., 1993; Kunkel et al., 1993), or with plants whose host-pathogen genetics is well known such as tomato (Martin et al., 1993; Salmeron et al., 1994).

Another key feature that will undoubtedly continue to be at the forefront of engineering resistance is to ensure that the resistance factor is made available at the site where it is best needed. Genes are now being inserted in potato, for example under the control of the 35S promoter of Cauliflower mosaic virus. Tissue specific promoters and sequences for signal peptides can be combined with "resistance" genes to give chimeric genes. These will ensure that gene products, e.g. lysozyme, will be produced in appropriate tissues and directed into cellular compartments or intercellular spaces where they are required.

Methods of insertion of genes of interest continue to be improved. For most current bacterial resistance work, transformation is usually based on *Agrobacterium tumefaciens* or *A. rhizogenes* but the risks of using this system may be more tightly regulated and researchers may choose alternative methods such as particle bombardment, or electroporation.

Durability of resistance is seen as a potential problem, more so for some methods than others. For example, a very minor mutation in a plantibody could render the resistance useless. R/avr gene resistance mechanisms could relatively easily be overcome. Now that engineering is more straightforward, it will be possible to incorporate two or more very different resistance mechanisms, for example a specific and a general mechanism. Durability would almost certainly be improved.

Resistance should not be considered an absolute requirement. Initial results show that tolerance is probably a more achievable goal, in which symptom expression is delayed, and severity is reduced. For diseases of quarantine significance such tolerance may not be so acceptable. Engineered disease resistance is likely to boom over the next decade and plant health authorities will almost certainly be under pressure from the international trade to review the status of certain diseases and the statutory regulations by which they are currently controlled.

One major area of work which must not be forgotten is the rigorous testing for *Agrobacterium* in engineered plant material. In the early period of release, it may be necessary for independent screening to be carried out rather than leave responsibility for this with the developers especially where they may be under commercial pressures. The seemingly relatively frequent accidental release of viable binary vector agrobacteria which appear to be free to exchange genetic information with wild type agrobacteria should prove a timely lesson.

One very recent critique of the gene revolution (Schmidt, 1995) indicates that no transgenic crops resistant to bacterial diseases are currently on their way to the market place. Of seven cases cited, three involve pest/disease resistance. Further commercial development may well depend on the success of these examples which, subject to USDA approval, will reach American markets in 1996.

9.4   CONCLUSIONS

Bacterial pathogens are probably less destructive to north-western European agricultural crops than are fungi, viruses or insects. Nevertheless they cause serious losses in both growing and stored potatoes, and these would certainly be larger, in the UK at least, if statutory regulations and inspections did not prevent the introduction of bacterial diseases that are not yet established there.

Research on the genetic manipulation of plants to produce resistance to bacterial pathogens has progressed far enough to indicate its feasibility. Some degree of resistance to specific bacteria has been induced by introducing R genes, and genes for enzymes resistant to bacterial toxins and genes for proteins toxic to bacteria derived from plants and other sources. The successful development of this work will need a better understanding of the natural processes of resistance, and of the expression and activity of introduced genes.

9.5   REFERENCES

Allefs, S.J.H.M., Florack, D.E.A., Hoogendoorn,C. and Stiekema, W.J. (1995). *Erwinia* soft rot resistance of potato cultivars transformed with a gene construct coding for antimicrobial peptide cecropin B is not altered. *American Potato Journal*, 72, 437-445.

Allefs, S.J.H.M. (1995) Resistance to *Erwinia* spp. in potato (*Solanum tuberosum* L.). PhD Thesis Landbouwuniversiteit Wageningen, Netherlands. 135 pp.

Carmona, M.J., Molina, A., Fernandez, J.A., Lopez-Fando, J.J., & Garcia-Olmedo, F. (1993). Expression of the alpha-thionin gene from barley in tobacco confers enhanced resistance to bacterial pathogens. *Plant Journal*, 3 457-462.

Chasan, R. (1994). Disease resistance: beyond the R genes. *Plant Cell*, **6**, 461-462.

Dangl, J., Debener, T., Dietrich, R.A., Gerwin, M., Kiedrowski, S., Ritter, C., Liedgens, H. & Lewald, J. (1993). Genetic resistance of *Arabidopsis* thaliana against phytopathogenic pseudomonads. [See Fuente *et al.*,] pp. 15-21.

Dangl, J.L., Ritter, L. Gibbon, M.J. Mur, L.A.J., Wood, J.R., Goss, S., Mansfield, J., Taylor, J.D. & Vivian, A. (1992). Functional homologs of the *Arabidopsis* RPM1 resistance gene in bean and pea. *Plant Cell*, **4**, 1359-1369.

Destefano-Beltran, L., Nagpala, P.G., Cetiner, S.M., Denny, T., & Jaynes, J.M. (1993). Using genes encoding proteins and peptides to enhance disease resistance in plants. In: *Biotechnology in disease control* (I. Chet ed.), pp. 175-189. Wiley-Liss Inc. New York.

Dons, J.J.M., Mollema, C., Stiekema, W.J. & Visser, B. (1991). Routes to the development of disease resistant ornamentals. In: *Genetics and breeding of ornamental species* (J. Harding, F. Singh and J.N.M. Mol eds.), pp. 387-417. Kluwer Academic, Netherlands .

During, K. (1993). Can lysozymes mediate antibacterial resistance in plants. *Plant Molecular Biology*, **23**, 209-214.

During, K., Fladung, M. & Lorz, H. 1993. Antibacterial resistance of transgenic potato plants producing T4 lysozyme. In: *Advances in molecular genetics of plant-microbe interactions* (E.W. Nester and D.P.S. Verma eds), pp. 573-577. Kluwer Academic, Netherlands.

Elphinstone, J.G. (1994). Inheritance of resistance to bacterial diseases. In *Potato genetics* (J.E. Bradshaw and G.R. Mackay eds), pp. 4429-446. Centre for Agriculture and Biosciences International (CABI) Wallingford, UK

EPPO (1994). Transgenic plants resistant to *Erwinia amylovora*. Report of Aldwinckle, H. (1994). Genetic engineering for disease resistance in apple. Paper presented at the 9th Congress of the Mediterranean Phytopathological Union, Turkey. EPPO Reporting Seminar 94/211. Issue Noá10 p 13.

Florack, D.E.A., Visser, B., De Vries, M., Van Vuurde, J.W.L., Stiekema, W.J. (1993). Analysis of the toxicity of purothionins and hordothionins for plant pathogenic bacteria. *Netherlands Journal of Plant Pathology*, **99**, 259-268.

Fuente, J.M. de la, Mosqueta, G. & Herrera-Estrella, L. (1993). Expression of a bacterial ornithyl transcarbamylase in transgenic plants confers resistance to phaseolotoxin. In: *Workshop on engineering plants against pests and pathogens, 11-13 January 1993, Madrid, Spain. Instituto Juan March de Estudios e Investigaciones, Madrid, Spain.* (G. Bruening, F. Garcia-Olmedo and F. Ponz eds), pp. 35-38.

Hiatt, A., Cafferkey, R. and Bowdish, K. (1989). Production of antibodies in transgenic plants. *Nature*, **342**, 76-78.

Hatziloukas, E. & Panapoulos, N.J. (1989). Nucleotide sequence and evolutionary origins of the two ornithine carbamoyl transferase genes in *pseudomonas syringae* pv *phaseolicola* (abstract). Fallen Leaf Lake Conference. Sept 14-16, 1989. p22.

Innes I.R.W., Bisgrove, S.R., Smith, N.M., Bent, A.F., Staskawicz, B.J. & Liu, Y.C. (1993). Identification of a disease locus in *Arabidopsis* that is functionally homologous to the RPGI locus of soybean. *Plant Journal*, **4**, 813-820.

Jaynes, J.M. (1993). Use of genes encoding novel lytic peptides and proteins that enhance microbial disease resistance in plants. *Acta Horticulture*, **336**, 33-38.

Jia, S.R., Xie, Y., Tang, T., Feng, L.X., Cao, D.S., Zhao, Y.L., Yuan, J., Bai, Y.Y., Jiang, C.X., Jaynes, J.M. & Dodds, J.D. (1993). Genetic engineering of Chinese potato cultivars by introducing polypeptide genes. *Current Plant Science and Biotechnology in Agriculture*, **15**, 208-212.

Kunkel, B.N., Bent, A.F., Dahlbeck, D., Innes, R.W. & Staskawicz, B.J. (1993). RPS2, an *Arabidopsis* disease resistance locus specifying recognition of *Pseudomonas syringae* strains expressing the avirulence gene avr Rpt 2. *Plant Cell*, **5**, 865-875.

Laflamme, D. & Roxby, R.W. (1989). Isolation and Nucleotide Sequence of cDNA clones of potato chitinase genes. *Plant Molecular Biology*, 249-250.

Martin, G., Vicente, C. de, Ganal, M., Miller, L. & Tankstey, S.D. (1993). Towards positional cloning of the Pto bacterial resistance locus from tomato. In: *Advances in molecular genetics of plant-microbe interactions* (E.W. Nester and D.P.S. Verma eds), pp. 451-455. Kluwer Academic, Netherlands.

Montanelli, C. & Nascari, G. (1991). Introduction of an antibacterial gene in potato (Solanum tuberosum L.) using a binary vector in *Agrobacterium* rhizogenes. *Journal of Genetics and Breeding*, **45**, 307-315.

Nascari, G., Montanelli, C., Chiara, T., Dodds, J.H. 1991. Genetic engineering of potato: An example of application of biotechnologies for the control of bacterial diseases in developing countries. *Revista di Agricoltura Subtropicale e tropicale*, **85**, 25-38.

Pirhonen, M. & Palva, E.T. (1988). Occurrence of bacteriophage T4 receptor in *Erwinia carotovora*. *Molecular and General Genetics*, **214**, 170-172.

Ronald, P.C., Salmeron, J.M., Carland, F.M. & Staskawicz, B.J. (1992). The cloned avirulence gene avr Pto induces disease resistance in tomato cultivars containing the Pto resistance gene. *Journal of Bacteriology*, **174**, 1604-1611.

Salmeron, J.M., Barker, S.J., Carland, F.M., Mehta, A.Y. & Staskawicz, B.J. (1994). Tomato mutants altered in bacterial disease resistance provide evidence for a new locus controlling pathogen recognition. *Plant Cell*, **6**, 511-520.

Schmidt, K. (1995). Whatever happened to the gene revolution. *New Scientist*, **145**, 21-25.

Shields, R. & Stratford, R. (1993). Debugging tomatoes. *Nature*, **366**, 508-509.

Taylor, J.D. Bevan, J.R., Crute, I.R. & Reader, S.L. (1989). Genetic relationship between races of *Pseudomonas syringae* pv. pisi and cultivars of *Pisum sativum*. *Plant Pathology*, **38**, 364-381.

Trudel, J., Potvin, C. & Asselin, A. (1992). Expression of active hen egg white lysozyme in transgenic tobacco. *Plant Science*, **87**, 55-67.

Turner, J.T., Kelly, J.L. & Carlson, P.S. (1993). Endophytes: an alternative genome for crop improvement. In *International Crop Science. I. International Crop Science Congress, Ames, Iowa, USA July 1992. Crop Science Society of America, Madison.* (D.R. Buxton, R. Shibles, R.A. Forsberg, B.L. Blad, K.H. Assay, G.M. Paulsen and R.F. Wilson eds), pp. 555-560.

Walter, B. (1991). Genie genetique applique a la vigne. *Bulletin de l'OIV*, **64**, 213-218.

Watanabe, K., Dodds, J.H., El-Nashaar, H., Zambrano, V., Benavides, J., Buitron, E., Siguenas, C., Panta, A., Salinsa, R., Vega, S. & Golmirazaic, A. (1993). Agrobacteirum-mediated transformation using antibacterial genes for improving resistance to bacterial wilt in tetrapoild potatoes. *American Potato Journal*, **70**, 853-854.

Yoneyama, X. & Anzai, H. (1993). Transgenic plants resistant to diseases by the detoxification of toxins. In: *Biotechnology in plant disease control* (I. Chet ed), pp. 115-137. Wiley-Liss Inc. New York.

10.   POTENTIAL MANIPULATION OF CROP RHIZOSPHERE

P.R. Hirsch
(IACR-Rothamsted)

10.1   BACKGROUND

That distinctive rhizosphere microflora are associated with particular crops have been known for many years (Phillips & Streit, 1994), although its significance is not clear. Different soil types, and the age of the plant have an effect. However, rhizodeposition from plants, particularly the soluble root exudates, must be the major factor which influences the soil microbiota (Lynch, 1990). It is well established that signal molecules in root exudate play an important part in the association between symbiotic rhizobia and their leguminous plant hosts, and there is some preliminary evidence that plant pathogens may be stimulated by specific components of root exudates (Nelson, 1990, Phillips & Streit, 1994). The signal molecules from susceptible plants that act as nematode hatching factors are another well-known example. Whether the normal commensal rhizoflora of a healthy plant is attracted by specific signals or by the distinctive combination of nutrients in exudates needs further research. In the plants that have been studied, the major components of exudate are sugars, organic acids, amino acids and vitamins, with lower levels of more unusual molecules such as phenolics and flavonoids which are potential signal molecules.

10.2   CROPS AND TARGETS

At present there is insufficient information on the basis of particular plant-soil pathogen interactions for detailed discussion on an individual basis. Crops and their major soil pathogens are listed elsewhere, therefore the topic will be treated as a general case.

10.3   PROSPECTS

Manipulation of plant physiology to alter root exudation could take two paths. Firstly, the synthesis and exudation of specific signal molecules which is known to attract pathogens and pests could be reduced or eliminated by genetic manipulation. However, this may not be feasible because signal compounds may also have a dual role and also attract beneficial organisms. The second approach would be to stimulate exudation of the factors that attract beneficial and commensal organisms that compete with pathogens.

Beneficial organisms include those that improve plant growth, e.g. though improving plant nutrition (rhizobia, Vesicular arbuscular mycorrhizal fungi), and those which stimulate or modify root growth by production of phytohormones (Gareth Jones, 1993). A wide variety of plant growth-promoting rhizobacteria (PGPRs) have been reported including rhizobia, azospirilla and pseudomonads (Schippers et al., 1990). Another group of beneficial organisms is those with biocontrol potential, although their efficacy in controlling major diseases in the field is often disputed. This may be because when they are applied to crops as seed or soil inoculants they do not survive well or compete with the native microflora to colonise roots, and do not have the efficacy demonstrated in

95

laboratory and glasshouse trials (Gareth Jones, 1993).

A well-known example is the ability of some rhizosphere fluorescent pseudomonads to produce phenazine antibiotics which strongly inhibit the take-all fungus of cereals, *Gaeumannomyces graminis* in laboratory culture (Thomashow & Weller, 1990). The application of such strains as biocontrol agents has not been proven in field trials and remains controversial. However, the phenomenon of take-all decline may be due to the build-up of bacterial populations with the ability to inhibit *G. graminis* in soils with long-term cereal cultivation. Fluorescent pseudomonads isolated from soil have also been shown to produce 2,4-diacetylphloroglucinol, an antibiotic that inhibits both damping-off fungi and *G. graminis* (Bangera *et al.*, 1994). Other pseudomonads appear to generate hydrogen cyanide which can control other microbes in the rhizosphere, and bacteria which secrete chitinases that have anti-fungal activity have also been described (Schippers *et al.*, 1990).

Other beneficial rhizosphere bacteria have been found to produce siderophores that chelate iron (Chet *et al.*, 1990). This is essential for microbial growth but is present in relatively low amounts in the rhizosphere. If the commensal rhizoflora removes iron from the rhizosphere, whether by siderophore production or by less efficient methods, it is not available for opportunist pathogens. This represents one mechanism by which the normal rhizoflora of a crop protect roots from pathogens by "niche exclusion". Vesicular arbuscular mycorrhizal fungi may play a similar role (in addition to improving phosphate nutrition) by colonizing and penetrating the regions of the root susceptible to infection, excluding other fungi. The protection may be purely physical, but there is some evidence that a degree of immunity to further fungal infection is induced in the plant (Gareth Jones *et al.*, 1993).

Seeds may also produce signal molecules or specific attractants during germination. Several volatile compounds from germinating seeds and root tips which stimulate the germination of pathogenic fungal spores have been described (Nelson 1990). Betaines such as trigonelline are found in many seeds where they protect against desiccation, and induction of rhizobial symbiotic genes by trigonelline has been shown (Phillips & Streit, 1994).

If mechanisms by which beneficial soil microbes are attracted to crop roots can be manipulated, inefficient inoculation of seeds or soil will be unnecessary. Instead, the indigenous microflora which is well-adapted to local soil conditions should provide a rhizoflora that protects the plants from soil pathogens.

10.4   CONCLUSIONS

Factors in root exudates influence the rhizoflora of different crops. Both beneficial and pathogenic organisms may be attracted by factors in exudate but if the roots are colonised initially by the numerous innocuous and beneficial soil microbes, opportunities for subsequent invasion by opportunist root pathogens will be reduced.

Very little is known at present about which specific factors are responsible for attracting soil microbes to roots. However, there is a possibility that plants could be manipulated to modify exudate composition, to enhance colonization by beneficial and innocuous microbes, or to cease to attract undesirable organisms. A prerequisite is identification of key exudate components which implies prior identification of innocuous and beneficial rhizosphere microorganisms which are competitive root colonisers.

Thus, there is an attractive long-term prospect of manipulating plants so that their root exudates create a more favourable rhizosphere either by not attracting pathogens, or by encouraging beneficial or competing microorganisms. But before this prospect is possible and the problems of genetic manipulation of plants are faced, much more basic information needs to be known about the microflora which characterize crops, the beneficial and damaging microorganisms identified, as well as the components of root exudates to which they respond.

10.5   REFERENCES

Bangera, M.G., Weller, D.M. & Thomashow, L.S. (1994). Genetic analysis of the 2,4-diacetylphloroglucinol biosynthetic locus from *Pseudomonas fluorescens* Q2-87. In *Advances in Molecular Genetics of Plant-Microbe Interactions Vol. 3*. Kluwer Academic Publishers, Dordrecht.

Chet, I., Ordentlich, A., Shapira, R. & Oppenheim, A. (1990). Mechanisms of biocontrol of soil-borne plant pathogens by rhizobacteria. *Plant and Soil*, **129**(1), 85-92

Gareth Jones, D. ed. (1993). Exploitation of Microorganisms. Chapman & Hall, London

Lynch, J.M. ed. (1990). The Rhizosphere. John Wiley, Chichester.

Nelson, E.B. (1990). Exudate molecules initiating fungal responses to seeds and roots. *Plant and Soil*, **129**(1), 61-73.

Phillips, D.A. & Streit, W. (1994). Legume signals to rhizobial symbionts: A new approach for defining rhizosphere colonization. In *Plant Microbe Interactions Vol. 1*. Chapman & Hall, N.Y.

Thomashow, L.S. & Weller, D.M. (1990). Role of antibiotics and siderophores in biocontrol of take-all disease of wheat. *Plant and Soil*, **129**, 93-99.

Schippers, B., Bakker, A.W., Bakker, P.A.H.M. & van Peer, R. (1990). Beneficial and deleterious effects of HCN-producing pseudomonads on rhizosphere interactions. *Plant and Soil*, **129**(1), 75-83.

11. SOME POTENTIAL RISKS AND HAZARDS ASSOCIATED WITH THE USE OF TRANSGENIC CROP PLANTS

P.R. Shewry and W.S. Pierpoint
(IACR-Long Ashton and IACR-Rothamsted)

1.   The commercial exploitation of pathogen-resistant, transformed plants, like that of any other transformed plants, depends of course upon their release into the environment being accepted both officially by government agencies and by the public. Environmental problems that could arise and which cause concern have been identified and discussed (see OECD, 1986) and, in the UK, are considered in the official (Department of the Environment) risk assessment procedure. Many experiments have been done, with rapeseed for example, concerning the vigour, survival and invasiveness of transformants (Crawley *et al.*, 1993), on the spread of genetically-tagged pollen in the field (Scheffler *et al.*, 1993), and on the survival of seed in field (non-agricultural) conditions and on cross pollination between rapeseed and wild species (Scheffler *et al.*, 1994). Some of these projects in the UK were funded jointly by industry, government (DTI) and former AFRC Institutes under an initiative (PROSAMO) which has now ended. It may be that enough scientific background is now known to justify the environmental release of such well studied crops. An alternative view is that this type of research should be continued, to examine, for example, the long term stability of modified and multiply modified genomes in the field. This work would help satisfy an acknowledged public and professional need for caution and surveillance, as well as exploit a scientifically unique situation.

2.   The use of parts of viral genomes to transform plants presents potential hazards, especially concerning their interaction with infecting viruses so as to enhance their pathogenicity (Section 4.5). Many virologists consider these risks small and manageable, but think that they should be addressed experimentally and that field releases should be monitored to allay official and public apprehension.

3.   Many strategies to produce pathogen-resistant plants involve the use of compounds that are toxic, notably proteins. Many such toxins are natural products that are already present in foodstuffs, such as thionins and ribosome-inactivating proteins. In addition their activity may be destroyed during food preparation and cooking. Nevertheless their use in engineered plants may be of concern in two respects. The first is their horizontal spread into non-target plants which are consumed in natural food chains. The second is that their presence at higher levels may lead to increased incidence of toxicity due to inadequate food preparation. It would therefore seem sensible to seek for or design toxins, such as the Bt toxins, which pose negligible risk to mammals.

4.   Although some plant proteins used to confer resistance may not be toxic when consumed by humans or livestock, their presence at high levels may lead to an allergenic response. For example, the 2S albumins of *Brassica* species which confer a degree of resistance to fungal infection (Terras *et al.*, 1993) are related to major seed allergens of mustard (Menéndez-Arias *et al.*, 1988), castor bean (Machado and Silva, 1992) and

Brazil nut (Melo *et al.*, 1994), and might therefore also be expected to lead to allergenic responses if incorporated into foodstuff. Any such proteins should therefore be fully evaluated in feeding tests before they are permitted to enter the food chain. Research to define and predict allergenicity would also be of value in saving much wasted effort.

5.   It is estimated that between 1986 and 1993, over 1000 field trials of transgenic plants, involving 31 different crop species, took place in 31 countries. Dale (1995) has recently summarised these and also contrasted the different national regulations which govern their conduct. He emphasises potential problems that could arise in the international trade of transgenic crops and their products, and the progress which is being made to achieve the "harmonized, equitable and responsible regulations" which will be needed to facilitate this trade.

REFERENCES

Crawley, M.J., Hails, R.S., Rees, M., Kohn, D. & Buxton, J. (1993). Ecology of transgenic oilseed rape in natural habitats. *Nature* **363**, 620-623.

Dale,P.J. (1995). R & D regulations and field trialling of transgenic crops. *Trends in Bio/Technology*, **13**, 398-403.

Machado, O.L.T. & Silva Jr. J.G. (1992). An allergenic 2S storage protein from *Ricinus communis* seeds which is a part of the 2S albumin precursor predicted by cDNA data. *Brazilian Journal of Medical and Biological Research*, **25**, 567-582.

Melo, V.M.M., Xavier-Filho, J., Lima, M.S. & Prouvost-Danon, A. (1994). Allergenicity and tolerance to proteins from Brazil nut (*Bertholletia excelsa* H.B.K.). *Food and Agricultural Immunology* **6**, 185-195.

Menéndez-Arias, L., Moneo, I., Domínguez, J. & Rodríguez, R. (1988). Primary structure of the major allergen of yellow mustard (*Sinapis alba* L.) seed, *Sin a* I. *European Journal of Biochemistry*, **177**, 159-166.

O.E.C.D. (1986). Recombinant DNA Safety Considerations. Organisation for Economic Co-operation and Development. pp 1-70 Paris. (ISBN 92-64-12857-3).

Scheffler, J.A., Parkinson, R. & Dale, P.J. (1993). Frequency and distance of pollen dispersal from transgenic oilseed rape (*Brassica napus*). *Transgenic Research*, **2**, 356-364.

Scheffler, J.A. & Dale, P.J. (1994). Opportunities for gene transfer from transgenic oilseed rape (*Brassica napa*) to related species. *Transgenic Research*, **3**, 263-278.

Terras, F.R.G., Schoofs, H.M.E., Thevissen, K., Osborn, R.W., Vanderleyden, J., Cammue, B.P.A. & Broekaert, W.F. (1993). Synergistic enhancement of the antifungal activity of wheat and barley thionins by radish and oilseed rape 2S albumins and by barley trypsin inhibitors. *Plant Physiology*, **103**, 1311-1319.

12.   GLOSSARY

These notes are intended to explain to non-specialists some technical terms *as they are used in the text*. They are not strict definitions

A-Amylase
A digestive enzyme that hydrolyses the amylose component of starch

Antisense RNA
RNA which is complementary to a specific messenger RNA, and will bind to it to prevent its functional expression

Chitinase
An enzyme that degrades chitin, an important structural carbohydrate of many fungal cell walls and insect exoskeletons

Coat protein
The structural, non-enzymic protein that encapsulates the nucleic acid of plant viruses

Co-suppression of genes
Introducing a foreign gene into a plant to augment the activity of a native gene, may result in the co-suppression of the activity of both native and introduced gene

β-1,3-Glucanase
An enzyme that hydrolyses specific bonds in glucans, which are carbohydrate polymers present in cell walls of plants and, especially, fungi

Helper proteins (or Helper-components)
Virus-produced proteins that are necessary for the transmission of viruses to other plants by insects, especially aphids

Hypersensitive response
The resistance response of a plant to a pathogen that produces a discrete, often necrotic lesion and only a limited spread of the pathogen

Lectin
One of a family of proteins that bind tightly to specific carbohydrate residues (e.g. mannose or N-acetyl glucosamine) in polymers such as glycoproteins, glycolipids and polysaccharides

Lysozyme
An enzyme that hydrolyses specific bonds in complex mucopolysaccharides including cell wall components, and will consequently destroy certain bacteria

Marker gene
A gene which can be introduced into another organism along with other genetic material and whose readily detected functioning indicates successful transformation

Particle gun
A device for introducing DNA into cells by bombarding tissue with DNA-coated metallic particles

Pathogenesis-Related Proteins
Proteins produced, often in large amounts, following infection. They include antifungal proteins such as chitinases

Phytoalexins
Defensive, low molecular-weight chemicals produced by some plants in response to pathogens

Plantibody
An antibody to a particular antigen, originally derived from an inoculated animal, which is produced in a transformed plant

Plasmid
An extrachromosomal, self-replicating piece of DNA, usually of bacterial origin, which can be isolated, modified and used to transport novel genetic material into the same or another organism

Polygalacturonase
An enzyme that hydrolyses and dissolves some components of the pectins of the primary cell walls of plants

Promoter
Region of a gene, 5' to the coding region, to which RNA polymerase attaches to initiate the transcription of the gene into messenger RNA.

Proteinase
An enzyme that hydrolyses the peptide bonds of proteins and is essential for their digestion

Protoplast
Plant or fungal cell from which the wall has been removed, usually by enzymic digestion

R-gene
A major dominant gene which confers resistance on a plant against a pathogen (or more usually a specific strain of the pathogen), and is thought to interact with a specific avirulence gene (avr) in the pathogen

Regeneration
The production of a whole plant, a regenerant, from a cultured cell or tissue, usually after this has been genetically modified *in vitro*

Regulatory sequence
Sequences of nucleotides, usually near the 5' end of a gene and in the proximity of the promoter, which regulate operation of the gene and so determine when and in what tissues it is expressed

Replicase (viral)
A viral encoded enzyme which replicates the viral nucleic acid

Rhizosphere
The region immediately surrounding a root in which there is chemical and physical interaction between the root and soil micro-organisms

Ribosome
A cytoplasmic particle on which protein is synthesized under the direction of strands of messenger RNA (translation)

Ribosome-inactivating protein
Proteins produced in some plants which inhibit ribosomal protein synthesis

Satellite nucleic acid
Pieces of nucleic acid which may accompany a specific plant virus, and on which they depend for replication.

Semiochemical
A volatile "signalling" compound produced by plants or insects which affects the behaviour of other insects and invertebrates

Thionins
A family of cysteine-rich peptides, initially isolated from cereals, which are toxic to many micro-organisms

Transformant
A transgenic organism whose genetic constitution has been modified by the introduction of foreign DNA

Transformation
Modifying the genetic constitution of an organism by introducing novel genetic material. This may code for a novel protein, or may modify or even prevent the expression of an endogenous gene

Transgene
A gene which has been transferred from one organism to another

Translation
The biochemical synthesis of a protein under the direction of messenger RNA and in which the sequence of amino acids in the protein is determined by the sequence of nucleotides in the RNA

Transposon or transposable element
A sequence of DNA which is able to leave one genomic locus and to become inserted into another; the so-called "jumping gene".

13.   ACKNOWLEDGEMENTS

We are grateful to the large number of scientists who gave us advice. Many of them completed and returned a questionnaire which helped to orientate the original report to MAFF. Some were especially helpful and generous with their time when we approached them later.   These include Drs. D. Baulcombe (JIC, Norwich), R. Cole (HRI Wellesbourne), I. Denholm (IACR-Rothamsted), G.A. Edwards (Shell Research Centre, Sittingbourne), D.J. Ellar (Cambridge), J.A. Gatehouse (Durham), P. Jarrett (HRI, Littlehampton), D.J. James (HRI, East Malling), J.D.G. Jones (JIC, Norwich), A.J. Maule (JIC, Norwich), T.P. Pirone (Lexington, USA) and G. Lomonossoff (JIC, Norwich). Dr. T.M.A. Wilson and his colleagues B.D. Harrison and H. Barker at SCRI (Dundee) were especially helpful with advice on virus research.

Mr. K.J.D. Hughes, now at the MAFF Central Science Laboratory, Harpenden, gave invaluable help with the original MAFF Report, which is gratefully acknowledged. Staff of IACR-Rothamsted who advised on the shape and conclusions of the final text of the report include Prof. B.J. Miflin, Dr. S. James, Prof. J.A. Pickett and Prof. R.T. Plumb. Mrs. Valerie Topps provided expert assistance in typing and formatting the report for publication.

Finally, we thank MAFF, and especially Drs. D. Shannon and J. King for funding the original report, and also for granting approval for its publication in this form.